Robert Etheridge, Robert L. Jack

The Geology and Palaeontology of Queensland and New Guinea

Robert Etheridge, Robert L. Jack

The Geology and Palaeontology of Queensland and New Guinea

ISBN/EAN: 9783337176013

Printed in Europe, USA, Canada, Australia, Japan

Cover: Foto ©ninafisch / pixelio.de

More available books at **www.hansebooks.com**

THE

GEOLOGY AND PALÆONTOLOGY

OF

QUEENSLAND

AND

NEW GUINEA.

BY

ROBERT L. JACK, F.G.S., F.R.G.S.,

GOVERNMENT GEOLOGIST FOR QUEENSLAND; FORMERLY OF THE GEOLOGICAL SURVEY OF SCOTLAND;

AND

ROBERT ETHERIDGE, JUNIOR,

GOVERNMENT PALÆONTOLOGIST (NEW SOUTH WALES); FORMERLY ASSISTANT-GEOLOGIST, GEOLOGICAL SURVEY OF
VICTORIA; ACTING-PALÆONTOLOGIST, GEOLOGICAL SURVEY OF SCOTLAND; AND SENIOR
ASSISTANT, DEPARTMENT OF GEOLOGY, BRITISH MUSEUM, LONDON.

PLATES AND MAP.

PUBLISHED UNDER THE AUTHORITY OF THE HON. W. O. HODGKINSON, M.L.A., F.R.G.S.,

Minister for Mines and Public Instruction, Queensland.

BRISBANE:
JAMES CHARLES BEAL, GOVERNMENT PRINTER, WILLIAM STREET.

LONDON:
DULAU AND CO., 37, SOHO SQUARE.

1892.

LIST OF PLATES.

EXPLANATION OF THE PALÆONTOLOGICAL PLATES.

Unless otherwise stated, the Figured Specimens are in the Geological Survey of Queensland Collection.

Other Collections are indicated by letters within brackets [·], as follows :—

British Museum ...	by	[B.M.]
Australian Museum	,,	[A.M.]
Queensland Museum	,,	[Q.M.]
Mining and Geological Museum, Sydney	,,	[M.G.M.]
Mr. C. W. De Vis	,,	[De V.]
The late Revd. W. B. Clarke	,,	[W.B.C.]*
The late Mr. R. Daintree	,,	[R.D.]†
Philosophical Society of Bath, England ...	,,	[P.S.B.]
The late Revd. J. E. Tenison Woods	,,	[J.E.T.W.]‡
Prof. H. A. Nicholson ...	,,	[H.A.N.]
Prof. A. Liversidge, &c.	,,	[A.L.]§

---

* Since burnt in the Garden Palace fire, Sydney.
Now partly in the British Museum, partly in the Queensland Museum, and partly lost sight of.
Now partly in the Mackay Museum, and partly in the Technological Museum, Sydney.
Now in the Macleay Museum, Sydney.

PLATE 1.

Fig. 1. *Stromatoporella*, sp. ind. Vertical section. ×12. Reid Limestone, Northern Railway.

Fig. 2. Do. . Tangential section. ×12.

Fig. 3. *Stromatopora*, sp. ind. Portion of a weathered surface, with button-shaped prominences. Reid Gap.

Fig. 4. Do. Vertical section of the same. ×12.

Fig. 5. Do. Tangential section of the same. ×12.

Fig. 6. *Heliolites porosa*, Goldf. A highly weathered fragment, with the calices, septa, and cœnenchyma beautifully shown. ×2. Reid Gap.

Fig. 7. *Heliolites Daintreei*, N. and E. A fragment showing the relative proportions and form of the corallites. Broken River Limestone. [B.M.]

Fig. 8. Do. Transverse section, with the calices, septa, and copious cœnenchyma. ×5.

Fig. 9. *Heliolites plasmoporoides*, N. and E. Transverse section, with calices and limited cœnenchymal tubuli. Broken River Limestone. [B.M.]

Fig. 10. Do. Portion of same. ×5.

Fig. 11. Do. Vertical section, with the Heliolite-like tabulæ in the tubuli. ×5.

Fig. 12. *Heliolites Nicholsoni*, Eth. fil. Transverse section, showing the large number of small cœnenchymal tubuli. ×4. Broken River Limestone. [B.M.]

Fig. 13. *Alveolites robustus*, Rominger? Portion of a bifurcating branch. Broken River Limestone. [B.M.]

Fig. 14. Do. Portion of the weathered surface. ×4.

Fig. 15. *Alveolites*, sp. ind. Several calices of a lobate or palmate form. ×6. Arthur's Creek. [B.M.]

Fig. 16. Do. Vertical section of a part of the corallum, showing the irregular walls and numerous tabulæ. ×6.

Fig. 17. Do. Horizontal section, exhibiting the thin walls of the corallites. ×6.

Fig. 18. *Romingeria? Foordii*, Eth. fil. Ramose corallum, with subalternate tubular corallites. ×1½. Reid Gap.

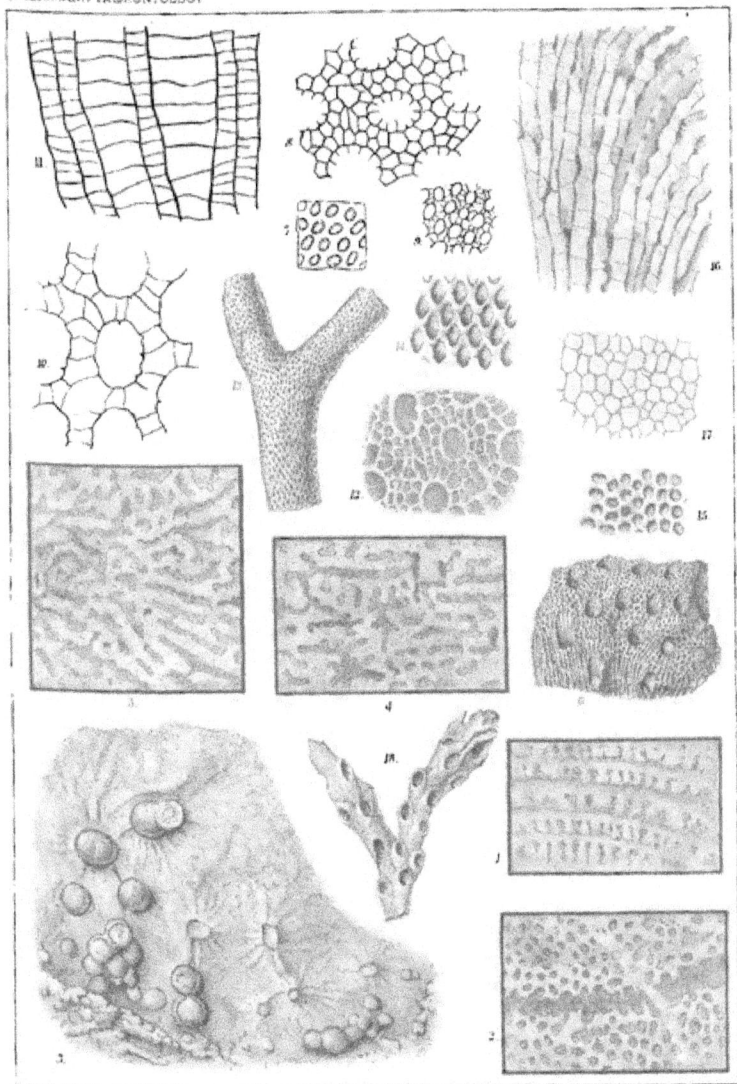

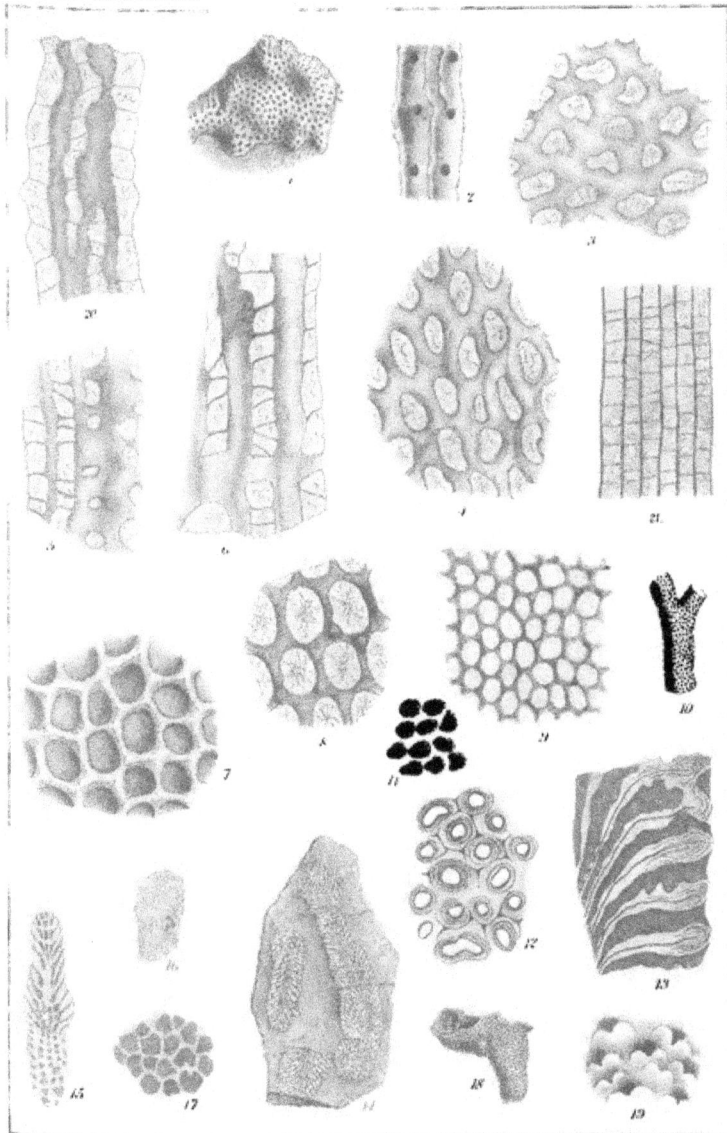

PLATE 3.

Fig. 1. *Favosites gothlandica*, var. *Goldfussi*, Ed. and H. Side view of a large weathered corallum. —⅓. Broken River. [B.M.]

Fig. 2. Do. do. Weathered surface, with a transverse view of the corallites.

Fig. 3. Do. do. Vertical or side view, showing the close-set tabulæ.

Fig. 4. *Favosites gothlandica*, Goldf. Upper weathered surface of a large flat laminar corallum. —⅓. Broken River.

Fig. 5. Do. Weathered surface, with transverse view of the corallites. ×6. [B.M.]

Fig. 6. *Aræopora australis*, N. and E. Transverse section taken from a polished surface, showing the form of the corallites and porous condition of the walls. ×2. Broken River. [B.M.]

Fig. 7. Do. Transverse section of the corallites, with the porous walls and trabecular septa. ×8.

Fig. 8. Do. Vertical section. ×2.

Fig. 9. Do. Vertical section, with the cribriform condition of the walls, septa, and rudimentary tabulæ. ×8.

Fig. 10. *Cyathophyllum* (comp. *C. helianthoides*, Goldf.). Transverse section of part of a corallite. Beaconsfield.

Fig. 11. *Cyathophyllum*, sp. ind. Exterior of the corallum, with growth accretion-ridges. Northern Railway.

Fig. 12. Do. Transverse section.

Fig. 13. *Cystiphyllum americanum*, var. *australasica*, Eth. fil. Side view of a partially decorticated example. Northern Railway.

Fig. 14. Do. Transverse section of the corallum.

Fig. 15. *Camphophyllum Gregorii*, Eth. fil. A natural section, vertical below, showing the denuded ends of the septa; oblique above, exposing irregular tabulæ. Northern Railway.

Fig. 16. Do. Transverse section of a smaller example, with marginal vesicular tissue, septa, and central tabulate area. Northern Railway.

Fig. 17. Do. Vertical section towards the base of the corallum, with slightly concave, irregular, or oblique tabulæ. ×1½. Northern Railway.

Fig. 18. Do. A similar example. ×1½.

Fig. 19. *Undescribed Turbinate Coral.* Broken River. ×2. [B.M.]

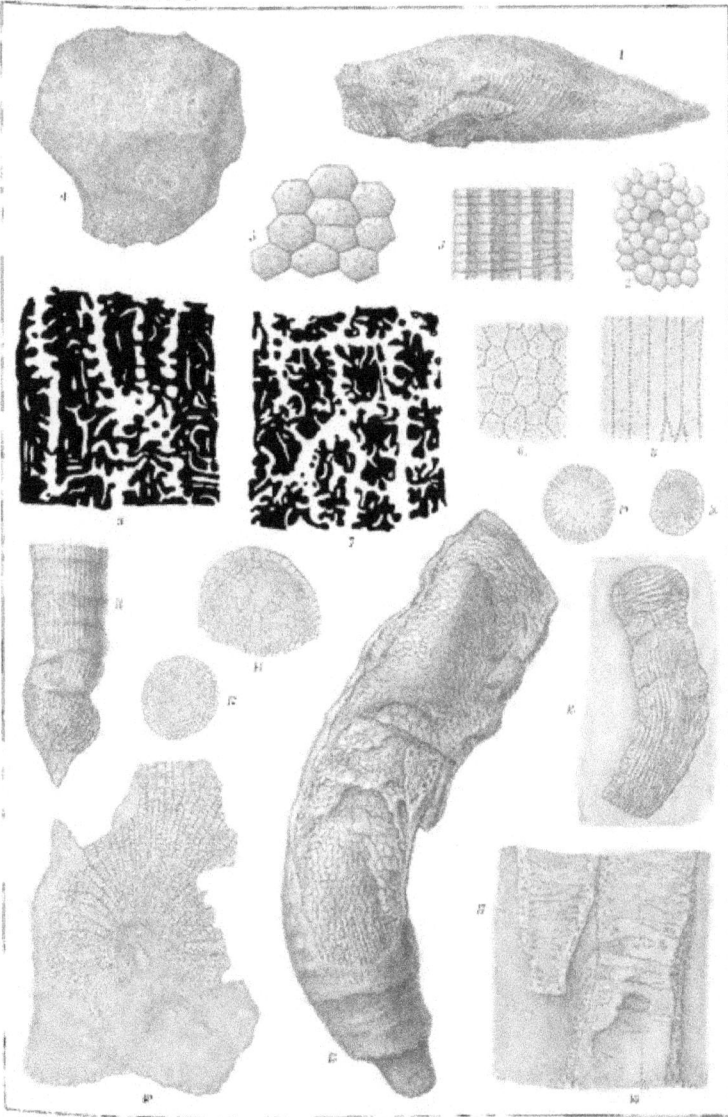

PLATE 4.

Fig. 1. *Spirifera euryglossa*, Schnur.? Ventral valve. Fanning Limestone.

Fig. 2. *Atrypa desquamata*, J. de C. Sby. Ventral valve, with radiating surface-ridges. Fanning Limestone.

Fig. 3. Do. Dorsal valve, with the same, and remains of the spirals. Fanning Limestone.

Fig. 4. *Atrypa reticularis*, Linn. Ventral valve. Fanning Limestone.

Fig. 5. *Rhynchonella primipilaris*, Von Buch. Ventral valve. Fanning River.

Fig. 6. *Gyroceras Philpi*, Eth. fil. Natural weathered surface, with septa and remains of the siphuncle. —¼. Northern Railway.

Fig. 7. Do. Portion of the external surface of the back, showing the blunt marginal tubercles, longitudinal grooves, and cross striation.

Fig. 8. *Lepidodendron Veltheimianum*, Sternb.? A decorticated example. Drummond Range.

Fig. 9. *Spondylostrobus*?, sp. ind. Vertical section of a large block. ×3. Wycarbah.

Fig. 10. Do. Horizontal section. ×3.

Fig. 11. *Asterocalamites scrobiculatus*, Schl. Two examples, mechanically united, one showing three and the other five internodes. Drummond Range.

Fig. 12. Do. Conical termination of the stem, with nine internodes. Drummond Range.

Fig. 13. *Dicranophyllum australienum*, Dawson. Upper Fanning Beds (after Dawson, *Quart. Journ. Geol. Soc.*, xxxvii., t. 13, f. 15).

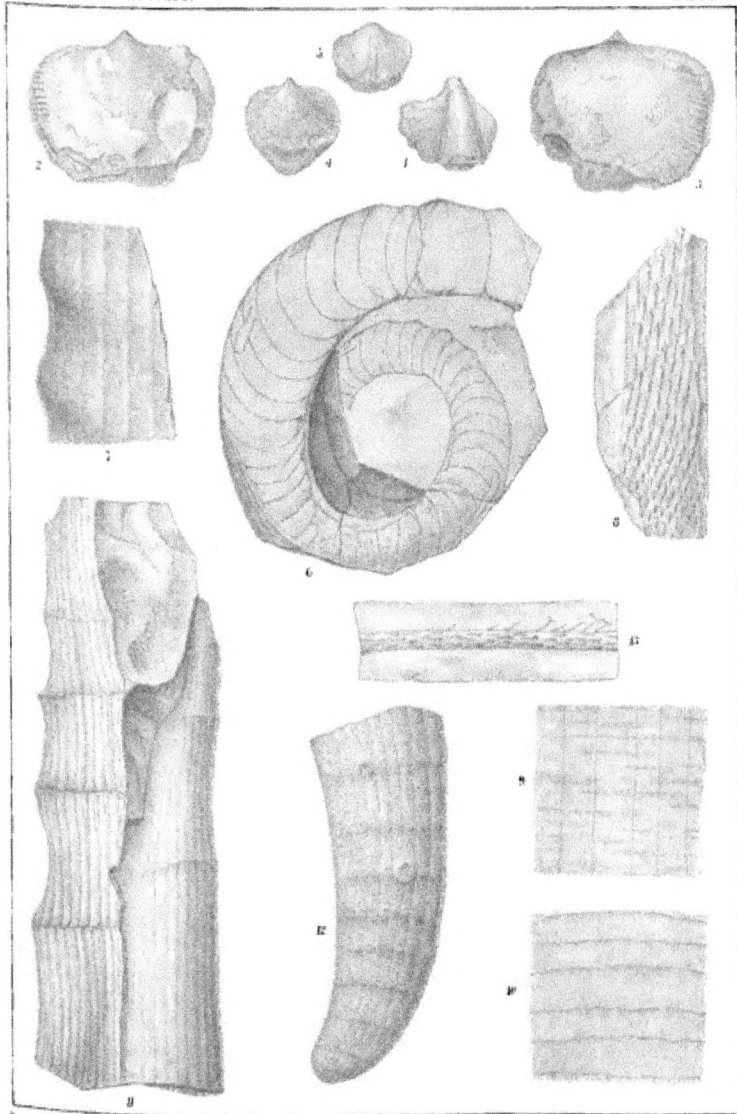

PLATE 4

PLATE 5.

Lepidodendron australe, McCoy.

Fig. 1. Part of a small branch with rhombic scars, showing the scar of the vascular-bundle as a line along the middle of each rhomb. Mount Wyatt or Canoona?

Fig. 2. Dichotomously dividing branch, the upper part clothed with leaves; the lower without leaves, and showing the rhombic scars of the latter. Mount Wyatt or Canoona?

Fig. 3. Portion of another branch, showing the passage of the rhombic leaf-scars into the more compressed scars of the upper part; the scar of the vascular bundle is situated on the upper angle of the leaf-scar. Mount Wyatt or Canoona?

Fig. 4. Part of a larger branch, with the vascular scar in the centre of the leaf-scar. Mount Wyatt or Canoona?

Fig. 5. Part of a stem, with large leaf-scars and apical vascular scar. Mount Wyatt or Canoona?

Fig. 6. A similar example. Drummond Range.

Fig. 7. Leaf-scars, without any vascular scar. Mount Wyatt or Canoona?

Fig. 8. Portion of a large branch, with the leaf-scars compressed laterally or perpendicularly. Drummond Range.

Fig. 9. Part of a branch, clothed with leaves. × 2½. Mount Wyatt or Canoona?

Fig. 10. Part of a somewhat older branch, deprived of the leaves, and showing the bases somewhat larger than in Fig. 1. × 2½. Mount Wyatt or Canoona?

[Figs. 1-5, 7, 9, and 10 are copied from the figures of Mr. Carruthers. He does not, in his Paper, distinguish between the localities of the specimens; they may, therefore, come either from Mount Wyatt or Canoona. The repository of these specimens, also, is unknown to me, unless they be in the Botanical Department of the British Museum.]

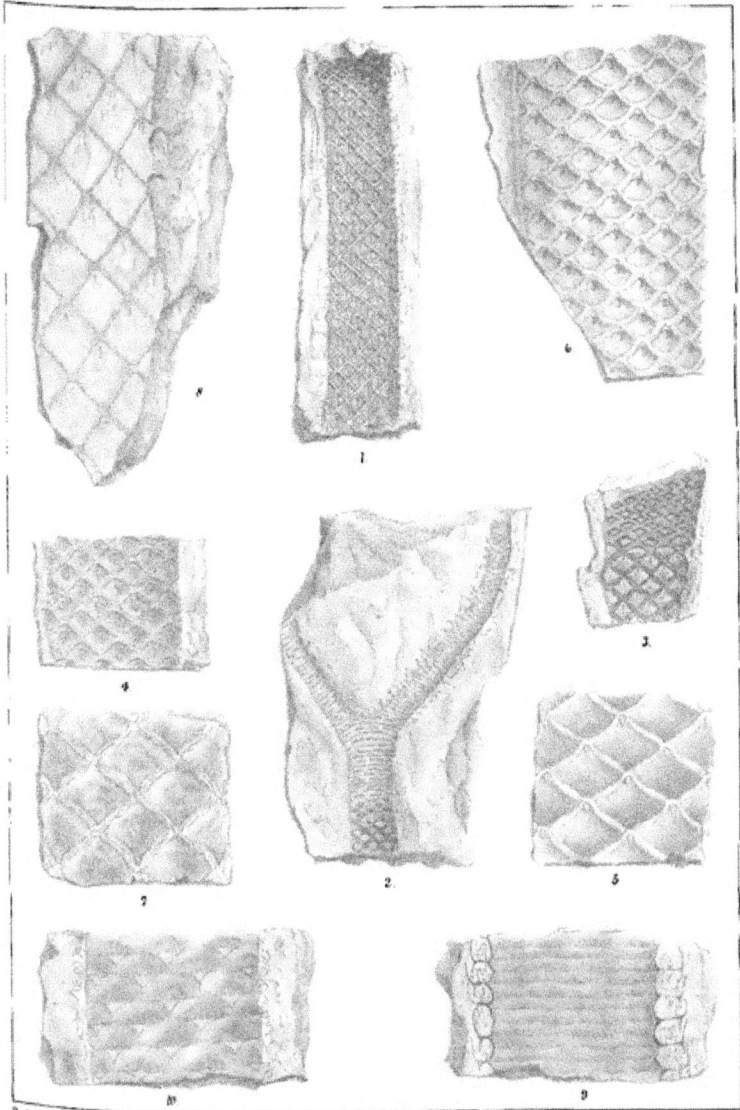

LOWER CARBONIFEROUS PLANTS

PLATE 6.

Fig. 1. *Lepidodendron*, sp. ind. Portion of a partially decorticated stem. (The scars are too distinctly drawn.) Drummond Range.

Fig. 2. *Knorria* condition of *Lepidodendron Veltheimianum*. Near Mount McConnell.

Fig. 3. Do. of Fig. 1? Drummond Range.

Fig. 4. *Lepidodendron Veltheimianum*, Sternb.? (The scars are too distinctly drawn.) Corner Creek, Star River.

Fig. 5. *Stenopora australis*, N. and E. Portion of a corallum, showing superimposed stage-growths. Bowen River. [B.M.?]

Fig. 6. Do. Three corallites, with moniliform expansions of the tubes; much enlarged. Bowen River.

Fig. 7. Do. Tangential section, with incomplete tabulæ and secondary infilling. ×24. Bowen River. [H.A.N.]

Fig. 8. Do. Vertical section of the corallites, with moniliform walls and tabulæ. ×24. Bowen River. [H.A.N.]

Fig. 9. *Stenopora Leichhardti*, N. and E. Tangential section, showing the form of the corallites, their thick walls, and numerous acanthopores. ×24. Bowen River. [H.A.N.]

Fig. 10. Do. Vertical section, exposing the walls of the tubes carrying acanthopores. ×24. Bowen River. [H.A.N.]

Fig. 11. *Stenopora? Jackii*, N. and E. Natural section of a branch. Bowen River.

Fig. 12. Do. Portion of the corallum, highly enlarged, displaying the curvature of the tubes and periodical thickenings. Bowen River.

Fig. 13. Do. A fragment, still more highly enlarged, showing supposed pores. Bowen River Coal Field.

Fig. 14. *Stenopora gimpiensis*, Eth. fil. Portion of a corallum fractured longitudinally, giving the characteristic appearance of the coral. Gympie.

Fig 15. Do. A few corallites, highly enlarged, to show angle of divergence, but the faint annulations are not represented.

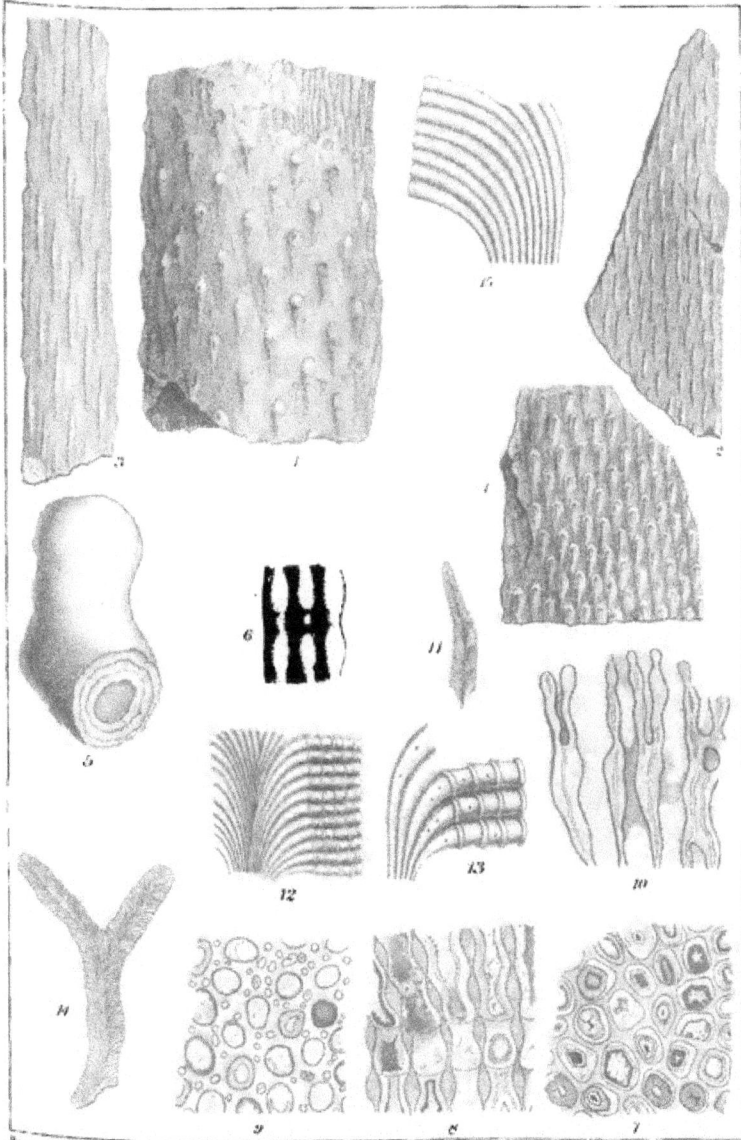

Bayeux & Highley del et lith   Mintern Bros. imp

LOWER CARBONIFEROUS PLANTS &

PLATE 7.

Fig. 1. *Cyathophyllum*, sp. ind. Horizontal natural section of an immense coral, which must have had a complete diameter of 7½ inches. Port Curtis.

Fig. 2. *Stenopora Leichhardti*, N. and E. Portion of a corallum, showing mode of growth, &c. × ½. Bowen River.

Fig. 3. *Orbipora? Waageni*. Portion of a bifurcating corallum, split in half naturally, showing the natural size of the corallites. Gympie.

Fig. 4. Do. Portion enlarged to show approximate size of the tubes and angle of divergence. × 2. Kooingal.

Fig. 5. Do. Transverse section exhibiting the form and relative size of the corallum. × 2. Kooingal.

Fig. 6. *Poteriocrinus crassus*, Miller? Impression of part of a stem with numerous cirri. − ¼. Stanwell.

Fig. 7. *Undescribed Crinoid*. Impression of eight arms, highly punctate. × 1½. Rockhampton.

Fig. 8. *Crinoid calyx and arms*. Portions of five arms and part of the calyx of an undescribed crinoid. × 2. Rockhampton.

Fig. 9. *Actinocrinus* or *Rhodocrinus*. Impressions of a few plates, highly ornamented with tubercules and radiating ridges. × 1½. Great Star River.

Fig. 10. *Granatocrinus? Wachsmuthi*, Eth. fil. Cast from the impression of a much crushed and a little distorted calyx, exhibiting one ambulacram and a deltoid plate. The first Blastoid found in Australia. × 1½.

Fig. 11. *Phillipsia Woodwardi*, Eth. fil. Portion of the glabella with thickened frontal margin, basal lobe, and three furrows on each side. × 1½. Stanwell.

Fig. 12. *Phillipsia dubia*, Eth. Thorax and pygidium, with part of the glabella (after Etheridge, *Quart. Journ. Geol. Soc.*, xxviii., t. 18, f. 7). Don River. [R.D.]

Fig. 13. *Phillipsia Woodwardi*, Eth. fil. Glabella without frontal margin. × 1½. Stanwell.

Fig. 14. *Griffithides seminiferus*, Phillips? Fragment of a free cheek, with a plain frontal border. × 4. Stanwell.

Fig. 15. *Beyrichia varicosa*, Jones. Internal cast of a right valve. × 8. Dotswood Reds.

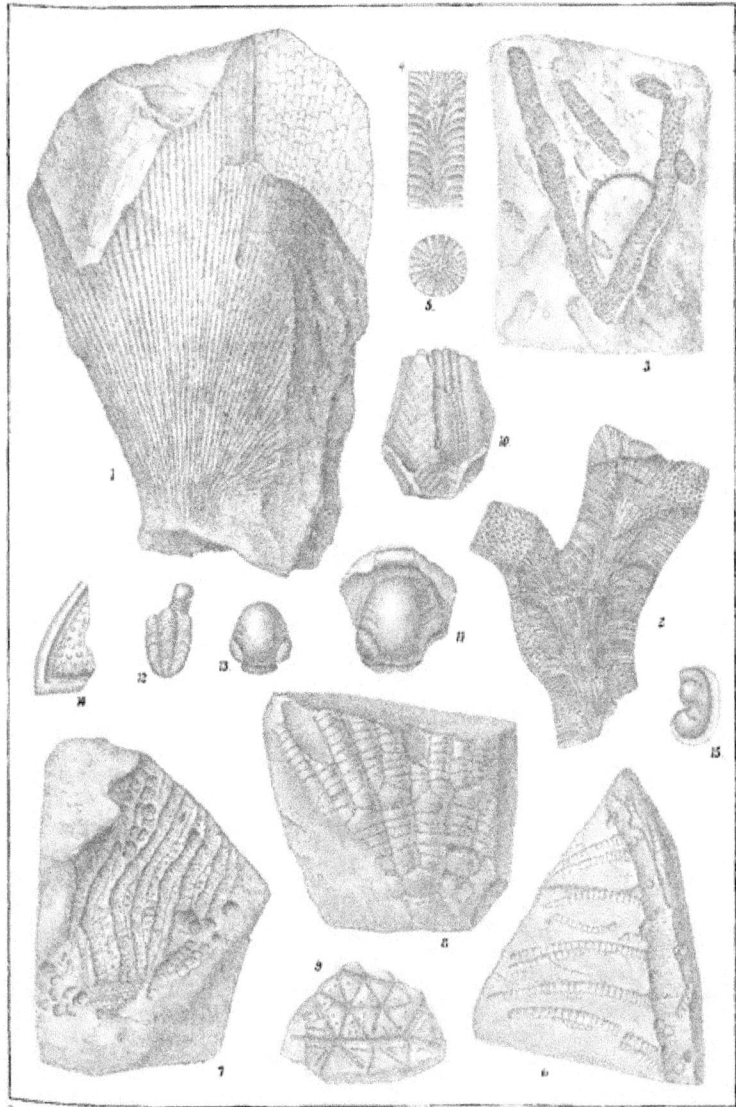

CARBONIFEROUS CORALS, ECHINODERMATA & CRUSTACEA

PLATE 8.

Fig. 1. *Poteriocrinus? Smithii*, Eth. fil.—Cast from an impression with stem and cirri, calyx, arms, and pinnulæ. ×1½. Stanwell.

Fig. 2. *Arenicolites*. Burrow at right angles to the bedding, infilled with matrix, and exposed by natural weathering. Stanwell.

Fig. 3. Do. Horizontal or surface view of another specimen, showing the clustered nature of the tubes and form of the aperture. Stanwell.

Fig. 4. *Track?*—formed of jointed cup-like segments. Athelstane Range, Rockhampton. (Loose on surface.)

Fig. 5. *Phillipsia dubia*, Eth. A decorticated and almost entire specimen, showing the peculiar outline of the glabella. (The pleuræ of the pygidium are too few by two or three.) ×3. Cornor Creek.

Fig. 6. Do. A larger pygidium, with the correct number of pleuræ on the pygidium. ×1½. Stanwell.

Fig. 7. *Fenestella multiporata*, de Koninck (*non* McCoy). Portion of a characteristic polyzoarium. Stanwell.

Fig. 8. Do. Portion much enlarged, with the basal plate removed, exposing the backs of the cells and the form of the fenestrules.

Fig. 9. *Fenestella* or *Polypora*. A carbonaceous filmy impression. Gympie.

Fig. 10. *Phyllopora*, or form allied to *Retepora laxa*, De Koninck. A film-like expansion of indefinite character. Gympie.

Fig. 11. Do. Portion enlarged, showing casts of cell-mouths on both the interstices and dissepiments.

Fig. 12. *Protoretepora ampla*, var. *Woodsi*, Eth. fil. An impression of a polyzoarium, with casts of the fenestrules. Bowen River.

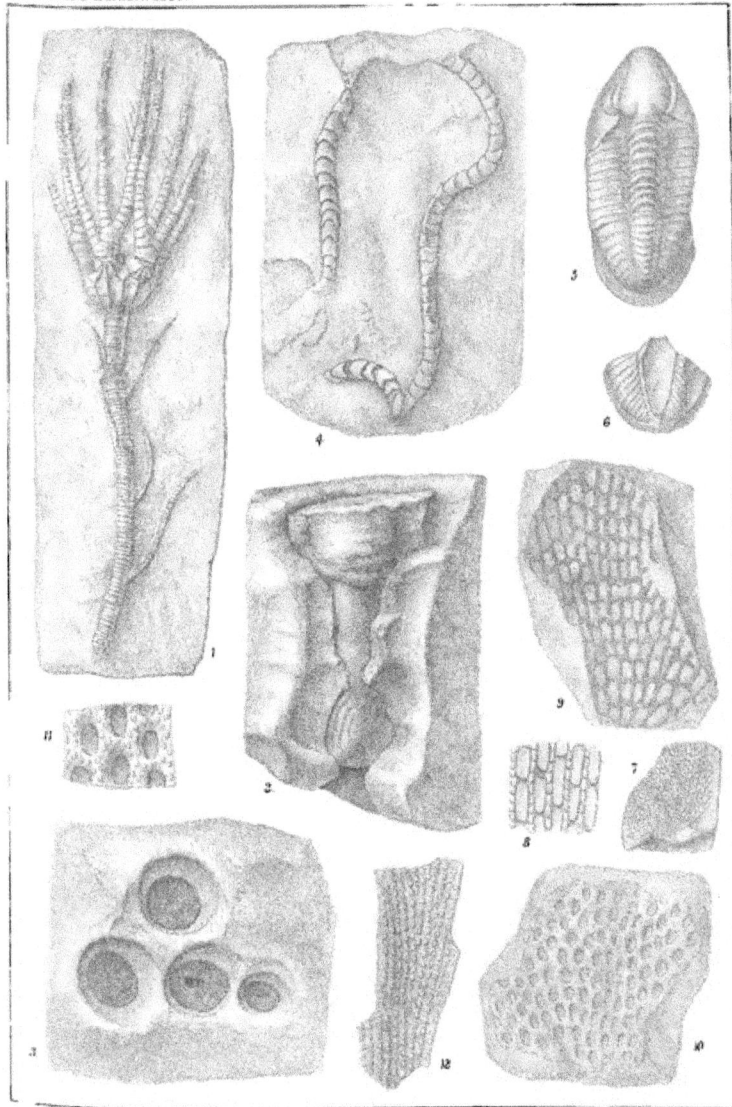

CARBONIFEROUS ANNELIDA, ECHINODERMATA, CRUSTACEA & POLYZOA.

PLATE 9.

Fig. 1. *Polypora Smithii*, Eth. fil. Impression of the back of portion of a polyzoarium. Stanwell.

Fig. 2. Do. Fragment highly magnified, showing the form of the fenestrules, and the striated interstices.

Fig. 3. Do. Impression of another example. Stanwell.

Fig. 4. *Fenestella fossula*, Lonsdale. Impression of the celluliferous face of a crumpled frond. Bowen River.

Fig. 5. Do. Fragment highly enlarged, with casts of the cell-mouths and fenestrules.

Fig. 6. *Fenestella internata*, Lonsdale. Impression of the poriferous face of a large colony. — $\frac{1}{1}$. Stanwell.

Fig. 7. Do. Portion highly magnified.

Fig. 8. *Rhombopora laxa*, Etheridge. Portion of a branch, with two bifurcations, and showing general habit. Gympie.

Fig. 9. Do. Natural section, with tubes enlarged. × 4.

Fig. 10. *Diclasma cymbæformis*, Morris, sp. Internal cast of a fine example. Bowen River.

Fig. 11. Do. Side view of the same specimen.

Fig. 12. *Spirifera*, allied to *S. striata*, Sby. Dorsal valve of a small individual (after Etheridge, *Quart. Journ. Geol. Soc.*, xxviii., t. 18, f. 8). Don River. [R.D.]

Fig. 13. *Spirifera Darwinii*, Morris. Ventral valve of a young specimen. Bowen River.

Fig. 14. Do. Dorsal valve.

Fig. 15. *Spirifera trigonalis*, var. *bisulcata*, Sby. Ventral valve (after Etheridge, *Quart. Journ. Geol. Soc.*, xxviii., t. 17, f. 4). Bowen River. [R.D.]

Fig. 16. Do. A large individual (*S. striata*, Etheridge, *Quart. Journ. Geol. Soc.*, xxviii., t. 17, f. 5. Bowen River.) [R.D.]

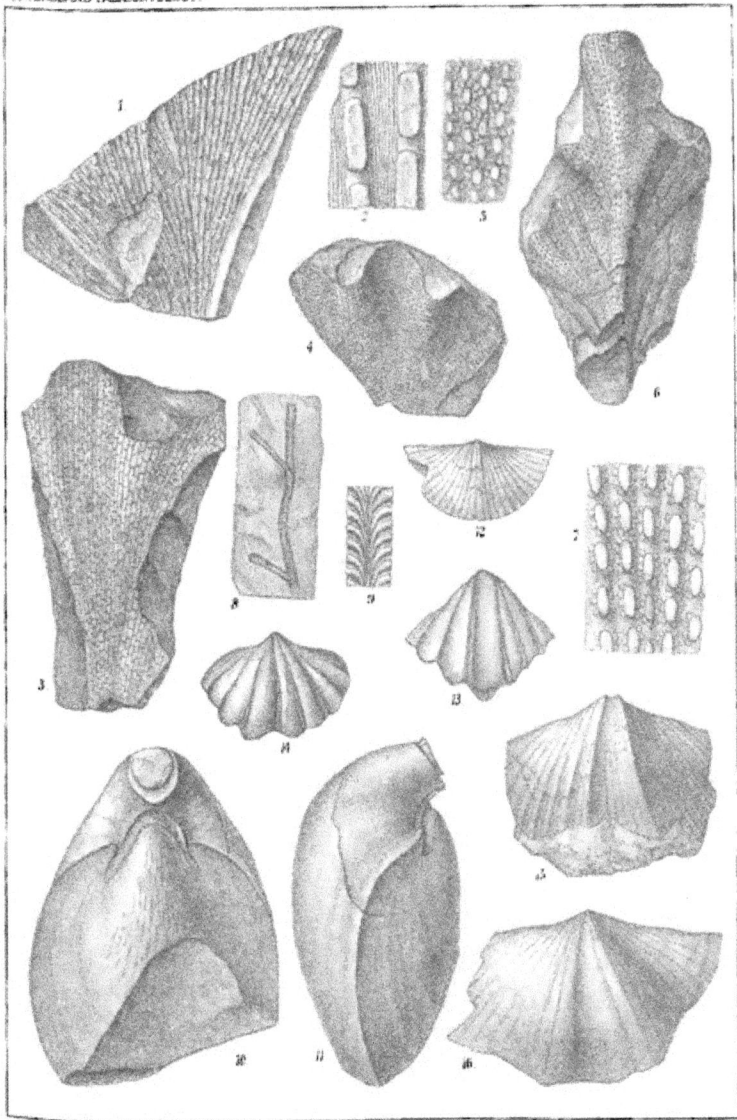

Berjeau & Highley del. et lith.   Mintern Bros. imp.

CARBONIFEROUS POLYZOA & BRACHIOPODA.

PLATE 10.

Fig. 1. *Spirifera tasmaniensis*, Morris? A decorticated specimen seen from the dorsal side, with the costæ much broken up. Bowen River Coal Field.

Fig. 2. *Spirifera Stokesii*, König. Internal cast of the dorsal valve, area, ventral umbo, &c. Mount Britton Gold Field.

Fig. 3. Do. The same specimen, seen from the ventral side.

Fig. 4. Do. Cast from an impression of a fragmentary specimen, showing the grouping of the radiating costæ. Gympie.

Fig. 5. *Spirifera Strzeleckii*, De Koninck. Cast of the exterior of a large dorsal valve, with simple widely separated costæ. Gympie.

Fig. 6. Do. Cast of the interior of the ventral valve (after Etheridge, *Quart. Journ. Geol. Soc.*, xxviii., t. 16, f. 3). Gympie. [R.D.]

Fig. 7. Do. (?) Cast of the interior of the ventral valve of a small example, slightly decorticated (after Etheridge, *Quart. Journ. Geol. Soc.*, xxviii., t. 15, f. 4). Gympie. [R.D.]

Fig. 8. *Spirifera vespertilio*, G. B. Sowerby. Cast of a dorsal (?) valve with a subdivided fold (after Etheridge, *Quart. Journ. Geol. Soc.*, xxviii., t. 16, f. 2). Gympie. [R.D.]

Fig. 9. *Spirifera bicarinata*, Eth. fil. Cast of a ventral valve, with the deep sinus and much extended hinge-line. Port Curtis.

Fig. 10. *Spirifera convoluta*, Phillips ? Cast of an indistinctly preserved ventral valve. Bowen River Coal Field.

Fig. 11. Do. Cast of the ventral valve of another example (after Etheridge, *Quart. Journ. Geol. Soc.*, xxviii., t. 17, f. 3). Bowen River. [R.D.]

Fig. 12. *Spirifera trigonalis*, Martin, var. *acuta*, Etheridge. Decorticated dorsal valve (after Etheridge, *Quart. Journ. Geol. Soc.*, xxviii., t. 16, f. 1). Gympie. [R.D.]

Fig. 13. *Spirifera bicarinata*, Eth. fil. Cast of a small ventral valve with a deep sinus. Stanwell.

Fig. 14. *Spirifera dubia*, Etheridge. Dorsal valve (after Etheridge, *Quart. Journ. Geol. Soc.*, xxviii., t. 16, f. 6). Gympie. [R.D.]

Fig. 15. *Spirifera*, sp. ind. Cast of a ventral valve, with costæ partially united in bundles.

Fig. 16. *Spirifera* (? *S. Clarkei*, De Kon.). Dorsal valve, with a sharp narrow fold, and equidistant simple costæ. Bowen River Coal Field.

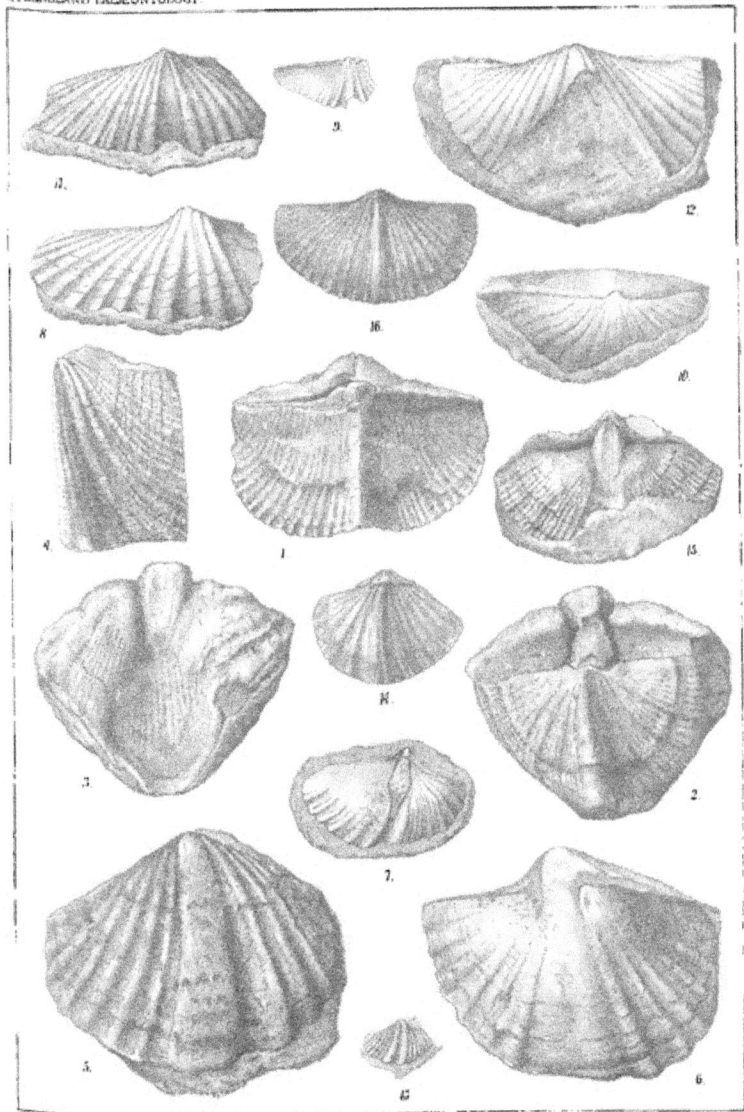

CARBONIFEROUS BRACHIOPODA

PLATE 11.

Fig. 1. *Spirifera bicarinata*, Eth. fil. Ventral valve. Great Star River Series.

Fig. 2. Do. Another specimen. Port Curtis District.

Fig. 3. Do. Portion of a ventral cast of a variety. Port Curtis District.

Fig. 4. *Spirifera duodecimcostata*, McCoy. Ventral valve. Cania.

Fig. 5. *Spirifera*, sp. ind. Impression of part of a large ventral valve, with numerous interpolated ribs. Stanwell.

Fig. 6. *Martinia? productoides*, Eth. fil. Ventral valve, possessing incipient radiating ribs in the middle line. Cania.

Fig. 7. Do. Another specimen, with a smooth surface.

Fig. 8. Do. Dorsal (?) valve of young form, with area and fissure.

Fig. 9. Do. Another similar specimen.·

Fig. 10. Do. Exterior of a young individual, but with a sulcus instead of a fold.

Fig. 11. Do. Ventral valve of another example, with area, fissure, and internal radiating ribs.

Fig. 12. *Martiniopsis? subradiata*, var. *Morrisii*, Eth. fil. Ventral valve of an abnormal variety. Cania.

Fig. 13. Do. Side view of the same specimen.

Fig. 14. *Martiniopsis? subradiata*, G. Sby., sp. Internal cast of the ventral valve. Stanwell.

Fig. 15. *Athyris Roysii*, Léveillé. Portion of an impression showing concentric imbrications, and the marginal radiating spines. (This is a very indifferent figure.) Stanwell.

Fig. 16. *Athyris Randsi*, Eth. fil. Dorsal view, showing the high fold. × 1½. Port Curtis.

Fig. 17. Do. Side view of the united valves. × 1½.

Fig. 18. Do. Front view of the same. × 1½.

Fig. 19. *Athyris ambigua*, Sby. A much crushed example, which is in consequence made to look too angular. (A bad figure.) Port Curtis.

Fig. 20. *Rhynchonella lilymerensis*, Eth. fil. Dorsal view. Lilymere.

Fig. 21. Do. Ventral view of the same specimen.

Fig. 22. Do. Side view of the united valves.

Fig. 23. *Rhynchonella pleurodon*, Phillips? Dorsal view. × 2. Cania.

Fig. 24. *Retzia radialis*, Phillips. Crushed example of the ventral (?) valve. Great Star River Series.

Fig. 25. Do. Portion of the surface, enlarged to show the casts of the shell perforations.

Fig. 26. *Orthis resupinata*, Martin. Decorticated ventral valve. Stanwell.

Fig. 27. *Orthis australis*, McCoy? Internal cast of the ventral valve, with the muscular impression casts well marked. Great Star River Series.

Fig. 28. *Orthis resupinata*, Martin. A small internal cast of the ventral valve referable to this species. Great Star River Series.

Fig. 29. *Orthis australis*, McCoy? A small example of the dorsal (?) valve. Stanwell.

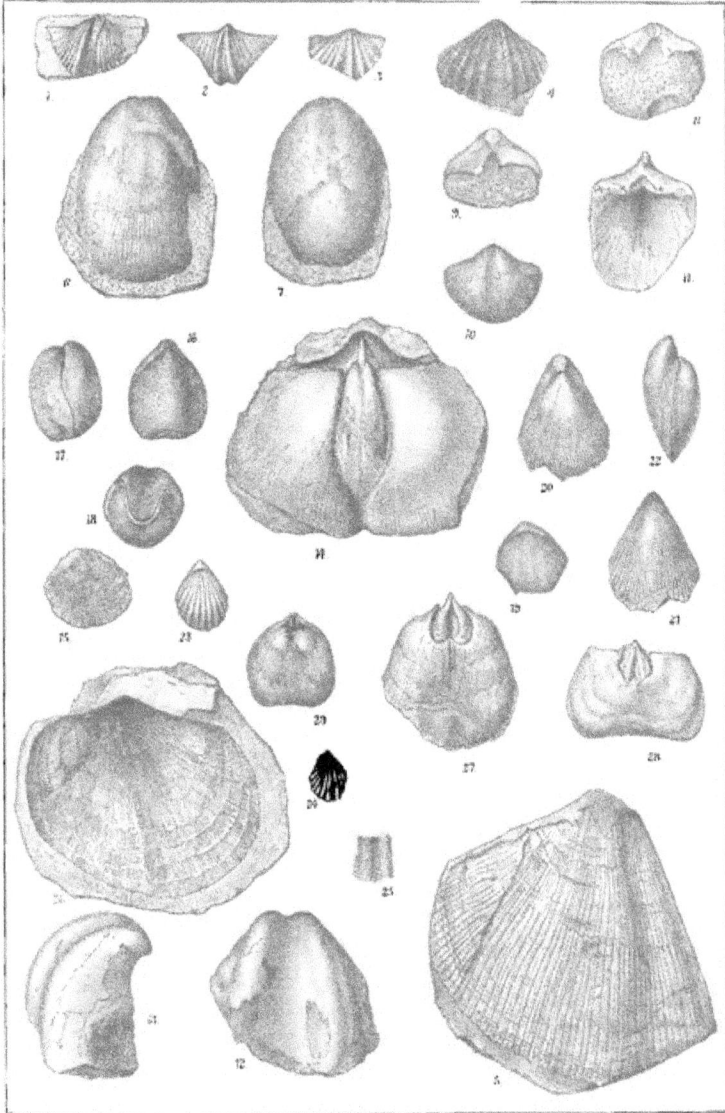

CARBONIFEROUS BRACHIOPODA.

PLATE 12.

Fig. 1. *Derbyia senilis*, Phillips, sp. Dorsal view. Bowen River Coal Field.

Fig. 2. Do. Side view of the same specimen.

Fig. 3. Do. Internal cast of a smaller example, showing the cavities left by the massive shelly plates diverging from the cardinal process. Bowen River Coal Field.

Fig. 4. Do. Fragment of the surface, highly enlarged to show the punctation of the test. ×5.

Fig. 5. Do. Portion of the area removed, exposing the bifid cardinal process of the dorsal valve. Bowen River Coal Field.

Fig. 6. Do. An undivided cardinal process.

Fig. 7. *Maccoyella reflecta*, Moore, sp. Cretaceous. (After Etheridge, as *Streptorhynchus Davidsoni*, *Quart. Journ. Geol. Soc.*, xxviii., t. 17, f. 1.) ? Walsh River. [Q.M.]

Fig. 8. *Strophomena analoga*, Phillips, sp. Dorsal valve (after Etheridge, *Quart. Journ. Geol. Soc.*, xxviii., t. 8, f. 1). Don River. [R.D.]

Fig. 9. Do. Dorsal valve (after Etheridge, *Quart. Journ. Geol. Soc.*, xxviii., t. 16, f. 7). Gympie. [R.D.]

Fig. 10. *Productus brachythærus*, G. Sowerby. Ventral view, showing long spine-bases. Bowen River Coal Field.

Fig. 11. Do. Dorsal view of the same specimen, with septum visible.

Fig. 12. Do. Lateral view, showing the somewhat geniculate form of the valve.

Fig. 13. Do. Lateral view of another example, with numerous impressions of spine-bases. Bowen River Coal Field.

Fig. 14. *Productus cora*, D'Orbigny. Ventral valve (after Etheridge, *Quart. Journ. Geol. Soc.*, xxviii., t. 15, f. 1). Gympie. [R.D.]

Fig. 15. *Productus*, sp. ind. (Compare *P. fimbriatus*, Sby.). Stanwell.

Fig. 16. *Productus undatus*, Defrance. Portion of the ventral valve of a small individual. Stanwell.

Fig. 17. *Productus*, sp. ind. Ventral valve, with very regular concentric laminæ. Burnett District.

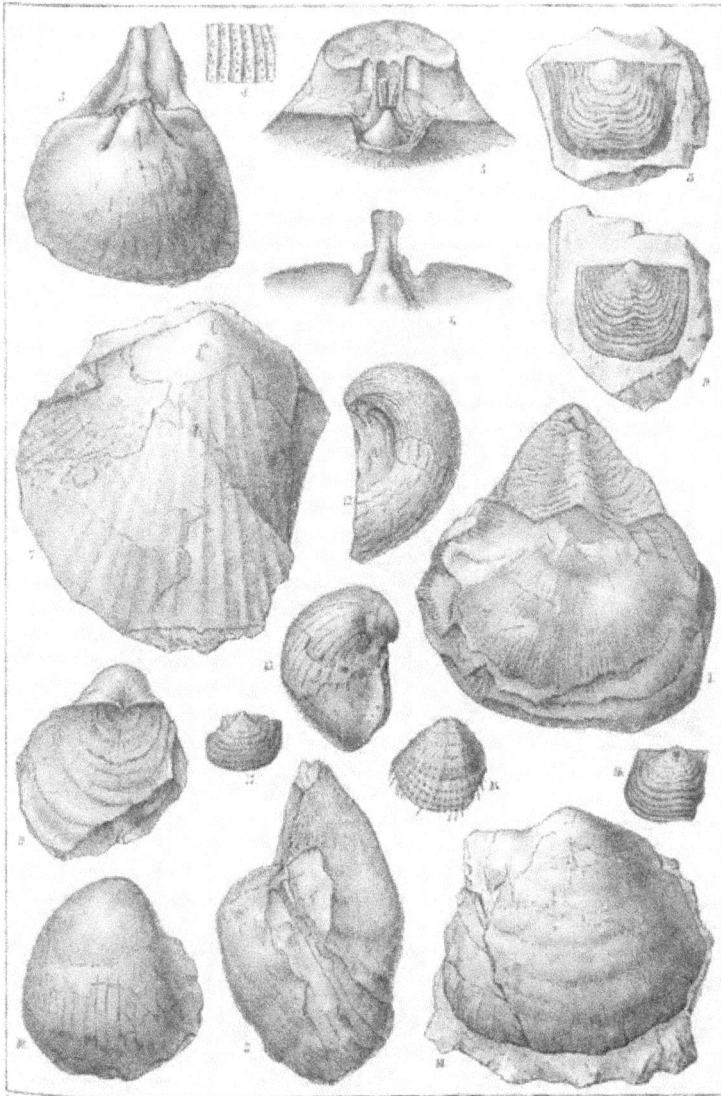

CARBONIFEROUS BRACHIOPODA.

PLATE 13.

Fig. 1. *Productus cora*, D'Orbigny. Ventral valve, decorticated. Stanwell.

Fig. 2. *Productus longispinus*, J. Sowerby? Dorsal valve (after Etheridge, *Quart. Journ Geol. Soc.*, xxviii., t. 18, f. 9). Don River.

Fig. 3. *Productus*, sp. ind. Partially decorticated ventral valve, exhibiting large muscular impressions, &c. (to compare with *P. brachythærus*, and *P. subquadratus*, t. 38, f. 7 and 8). N.S. Wales.

Fig. 4. *Productus*, sp. ind. Cast of a decorticated ventral valve, with prominent spine bases in the median sinus, after the type of *P. prælongus*, Davidson. Stanwell.

Fig. 5. *Productus brachythærus*, G. Sowerby? Ventral valve, much decorticated and somewhat crushed, exhibiting decurrent spine-bases. Stanwell.

Fig. 6. *Productus*, sp. ind. Decorticated ventral valve, with a sharp umbo, and fine, distinct, string-like costæ, after the type of *P. striatus*, Fischer. Burnett District.

Fig. 7. *Chonetes*, sp. ind. Dorsal valve, with exceedingly numerous bifurcating costæ. (? = Figs. 8 and 11.) × 1½.

Fig. 8. *Chonetes*, sp. ind. Dorsal valve, much decorticated and highly punctate (? = Figs. 7 and 11). × 1½. Athelstane Range.

Fig. 9. *Chonetes cracowensis*, Etheridge. Dorsal valve. Great Star River Beds.

Fig. 10. *Chonetes* (deltoid species). Ventral valve. × 1½. Great Star River Beds.

Fig. 11. *Chonetes*, sp. ind. Dorsal valve, internal cast (? = Figs. 7 and 8). Stanwell.

Fig. 12. *Strophalosia Clarkei*, Etheridge. Ventral valve, partly an internal cast, partly with the test preserved. Bowen River Coal Field.

Fig. 13. Do. Dorsal valve of another specimen, with the test preserved (after Etheridge, *Quart. Journa. Geol. Soc.*, xxviii., t. 17, f. 2 b). Bowen River Coal Field. [*R.D.*]

Fig. 14. Do. Another specimen (after Etheridge, *Quart. Journ. Geol. Soc.*, xxviii., t. 18, f. 4). Nogoa River. [*R.D.*]

Fig. 15. Do. Interior of the dorsal valve, showing dental sockets, adductor, and reniform impressions, septum, &c. Bowen River Coal Field.

Fig. 16. Do. Interior cast of the ventral valve, showing the impressions of the cardinal and adductor muscles, channels of spine-bases, septum, &c. Bowen River Coal Field.

Fig. 17. Do. Internal cast of the ventral valve of another specimen, with the dental sockets, and adductor and cardinal muscular impressions visible. Bowen River Coal Field.

Fig. 18. *Strophalosia Gerardi*, King? Dorsal view, displaying concave dorsal valve, area, and umbo of ventral valve. Bowen River Coal Field.

Fig. 19. *Lingula mytiloides*, J. Sowerby (or *L. ovata*, Dana?) Ventral valve. × 2. Cania District.

PLATE 13.

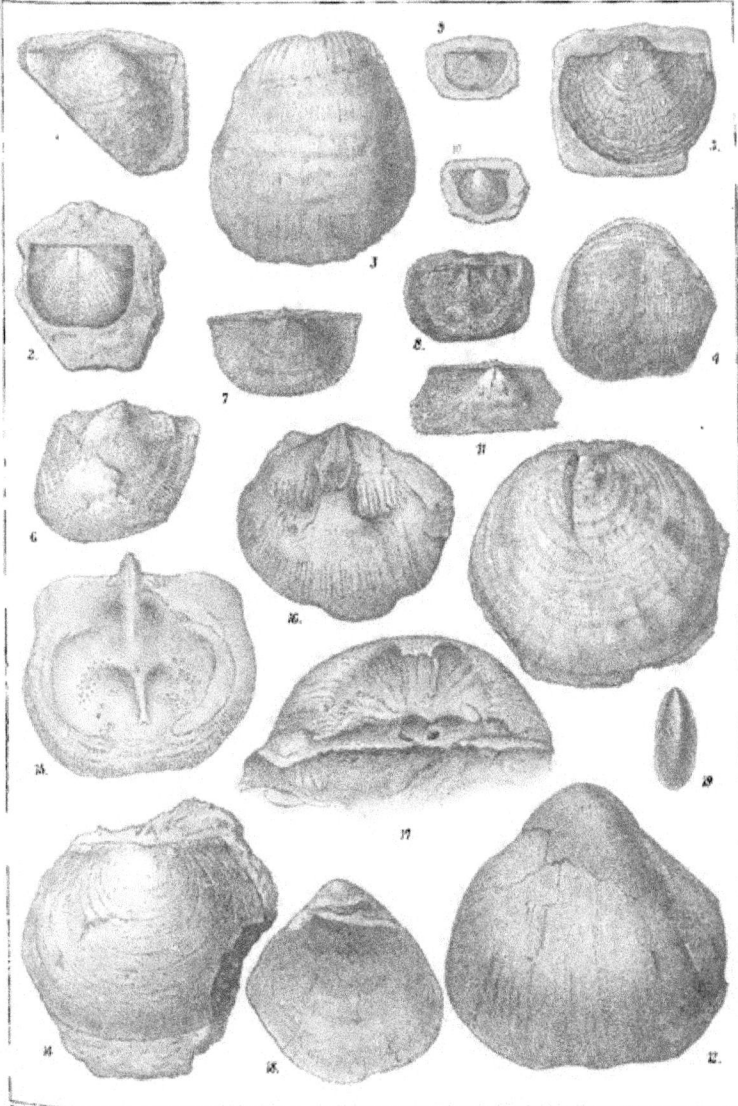

PLATE 14.

Fig. 1. *Aviculopecten limæformis*, Morris? Anterior ear of a large valve (the remainder of the specimen is not figured). Gympie.

Fig. 2. *Pachydomus*, sp. A badly preserved and somewhat mutilated right valve. ½ nat. Gympie.

Fig. 3. *Astartila cytherea*, Dana. Bowen River Coal Field.

Fig. 4. Do. View of the hinge-line of the same specimen.

Fig. 5. *Modiomorpha? mytiliformis*, Eth. fil. An imperfect left valve, with the umbo and anterior end removed. Banana Creek.

Fig. 6. *Conocardium australe*, McCoy. A decorticated example. Gympie.

Fig. 7. *Goniatites*, sp. ind. A partially decorticated example, believed to belong to this genus. Port Curtis.

Fig. 8. *Euomphaloid Shell.* Cast of a depressed undetermined shell, with close fine ornament. Stanwell.

Fig. 9. *Pleurotomaria* or *Mourlonia?* Decorticated specimen. Burnett District.

Fig. 10. *Entolium*, sp. ind. A decorticated example. Great Star River.

Fig. 11. *Cypricardella Jackii*, Eth. fil. Left valve with fine sculpture. × 1½. Mount Hamilton, near Gympie.

Fig. 12. Do. Hinge-line of the same shell. × 1½.

Fig. 13. *Modiomorpha? Daintreei*, Eth. fil. Right valve decorticated. Port Curtis.

Fig. 14. *Pleurophorus Randsi*, Eth. fil. Right valve with median diagonal costæ. Burnett District.

Fig. 15. *Astartella rhomboidea*, Eth. fil. Right valve with characteristic sculpture. Gympie. (This figure has been drawn rather out of position—the hinge line should be parallel to the top of the plate.)

Fig. 16. *Solenya Edelfelti*, Eth. fil. A highly radiate species. Bowen River Coal Field.

Fig. 17. *Nuculana*, sp. ind. A left valve, decorticated, closely resembling an American species. × 2. Great Star River.

Fig. 18. *Strophalosia Gerardi*, King. Exterior of the ventral valves. Bowen River Coal Field.

Fig. 19. *Strophalosia Clarkei*, Etheridge. Internal cast of the ventral valve. Bowen River Coal Field.

Fig. 20. *Myilops*, sp. ind. A decorticated example which should be compared with *M.? Bigsbii*, de Kon. Gympie.

Fig. 21. *Euomphaloid Shell.* A cast, rather obliquely pressed out of shape. × 2. Great Star River.

Fig. 22. *Naticopsis* or *Platyschisma?* Decorticated example seen from the anterior. Great Star River Series.

Fig. 23. Do. The same seen from the posterior.

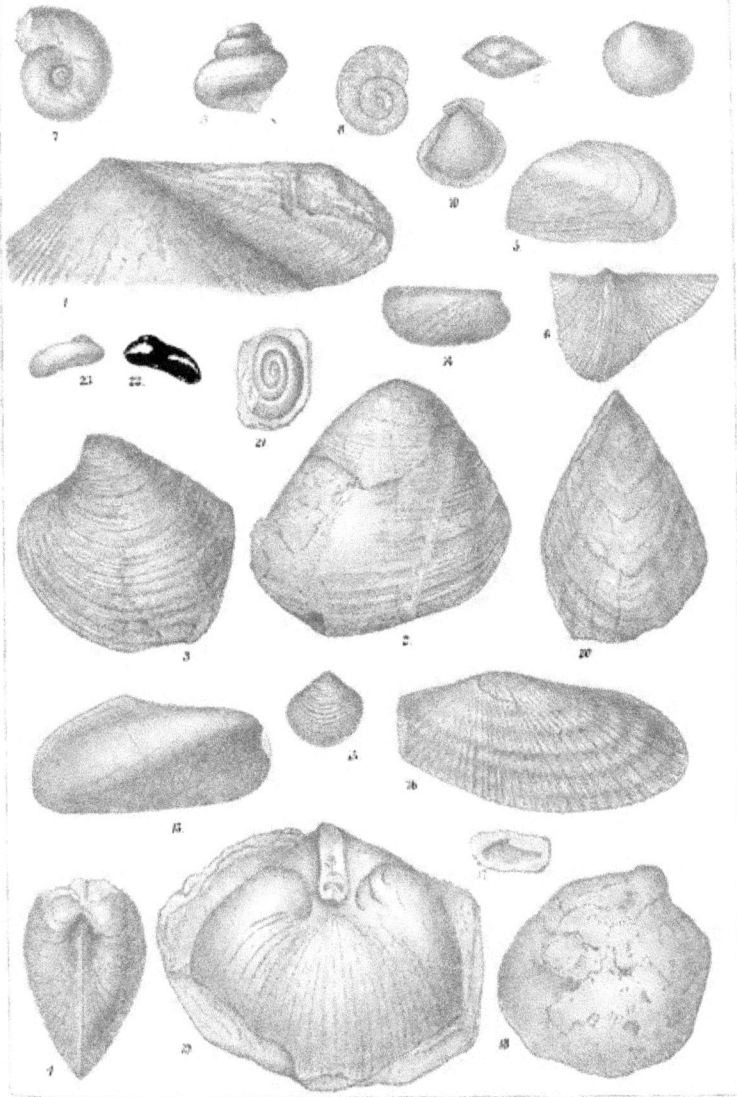

CARBONIFEROUS BRACHIOPODA & PELECYPODA

PLATE 15.

Fig. 1. *Orthoceras*, sp. ind. Showing traces of a lateral siphuncle. Gympie.

Fig. 2. *Maurlonia Strzeleckiana*, Morris, sp. Cast with faint indications of the band.

Fig. 3. *Platyschisma oculus*, J. de C. Sowerby, sp. A crushed and somewhat distorted example. Stanwell.

Fig. 4. Do. Side front view of the same specimen.

Fig. 5. *Goniatites?* sp. ind. A narrow, undetermined form. Gympie.

Fig. 6. *Platyschisma rotundata*, Etheridge. Anterior view, with aperture (after Etheridge, *Quart. Journ. Geol. Soc.*, xxviii., t. 18, f. 3). Cracow Creek. [R.D.]

Fig. 7. *Porcellia Pearsi*, Eth. fil. Imperfect lateral impression. ×2. Great Star River Series.

Fig. 8. Do. Back with spiral striæ, and elongated tubercles. ×2.

Fig. 9. *Murchisonia carinata*, Etheridge. Posterior view of four whorls (after Etheridge, *Quart. Journ. Geol. Soc.*, xxviii., t. 18, f. 5). Don River. [R.D.]

Fig. 10. *Naticopsis harpæformis*, Etheridge. Fragment of a whorl (after Etheridge, *Quart. Journ. Geol. Soc.*, xxviii., t. 18, f. 6). [R.D.]

Fig. 11. *Bellerophon stanwellensis*, Eth. fil. Anterior view with portion of aperture, reflected inner lip, and alar expansion. Stanwell.

Fig. 12. Do. Side view.

Fig. 13. Do. Posterior view.

Fig. 14. *Goniatites*, sp. ind. Fractured specimen showing inner whorls, and septa Rockhampton District.

Fig. 15. Do. Part of back of the same specimen, with spiral striæ, and growth-varex. Rockhampton.

Fig. 16. *Pleurotomaria carinata*, Etheridge. An internal cast, probably distorted (after Etheridge, *Quart. Journ. Geol. Soc.*, xxviii., t. 15, f. 6). Gympie. [R.D.]

Fig. 17. *Loxonema*, sp. ind. Posterior view. Stanwell.

PLATE 25

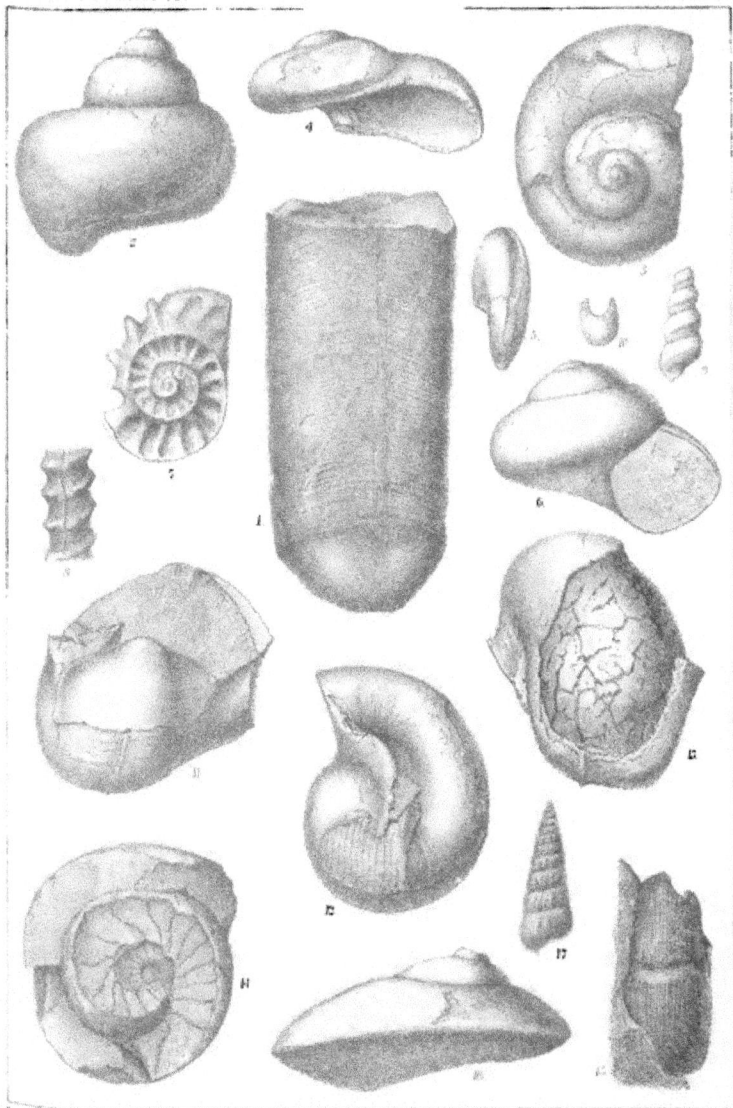

CARBONIFEROUS GASTEROPODA & CEPHALOPODA.

PLATE 16.

Fig. 1. *Alethopteris australis*, Morris, sp. Portion of a small pinna. Ipswich Coal
Measures, Colinton.

Fig. 2. *Ptilophyllum oligoneurum*, Ten. Woods ? Part of a small frond. Burrum
Coal Measures.

Fig. 3. *Pterophyllum*, sp. ind. (Compare *P. Carterianum*, O. and M.) Ipswich Coal
Measures, Colinton.

Fig. 4. *Tæniopteris*, sp. ind. Portion of a large frond, at present undetermined.
(The anastomosis of the venules is an error.) Mount Esk.

Fig. 5. *Macrotæniopteris crassinervis*, Feistmantel ?. Apex of a large frond. Wycarbah.

Fig. 6. *Glossopteris*, sp. ind. Portion of a finely-grained frond. Near Townsville.

Fig. 7. *Glossopteris ampla*, Dana. Part of a frond. Near Townsville.

Fig. 8. *Glossopteris Browniana*, Brongniart. Portion of a frond. Near Townsville.

Fig. 9. *Equisetaceous Phragma*. Stewart's Creek, Rockhampton. × 2.

Fig. 10. *Cycadinocarpus ?*, sp. ind. A Cycadaceous seed. Wycarbah.

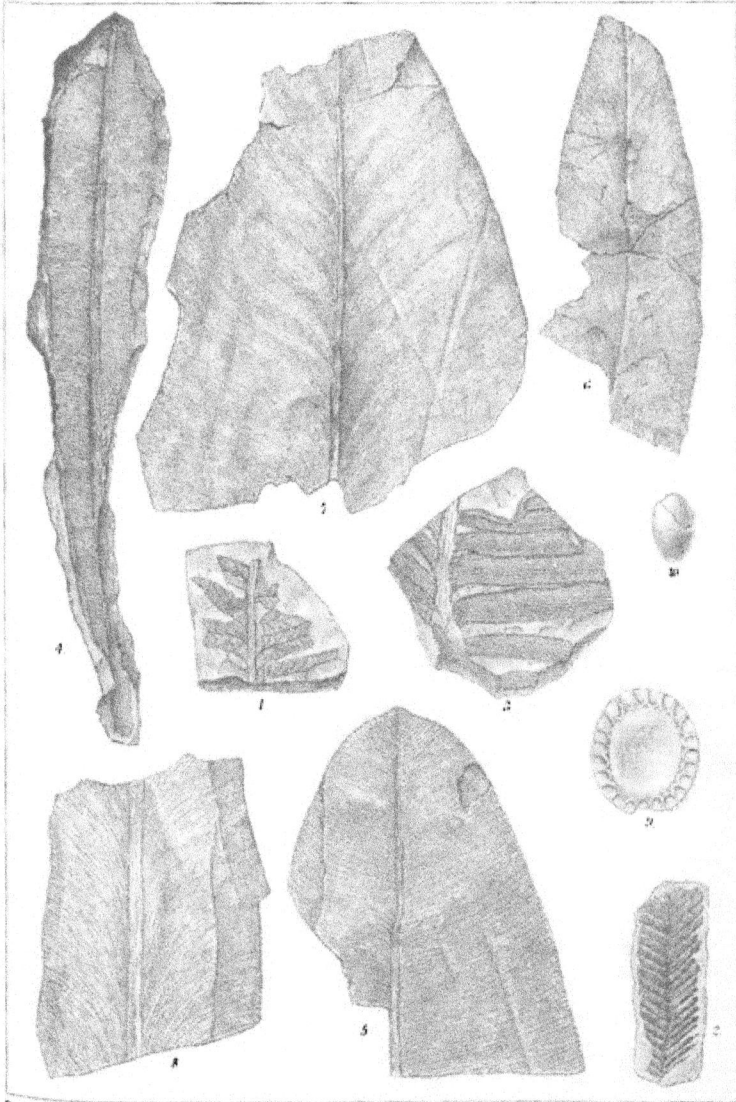

CARBONIFEROUS PLANTS.

PLATE 17.

Fig. 1. *Thinnfeldia odontopteroides*, Morris, sp. Terminal portion of a pinna. Ipswich Coal Measures.

Fig. 2. *Thinnfeldia media*, Ten. Woods ?, or *T. indica*, Feistmantel ? Burrum Coal Measures.

Fig. 3. *Alethopteris Lindleyana* (Royle), Feistmantel ? Mount Esk.

Fig. 4. Do. Pinnule enlarged.

Fig. 5. *Pterophyllum abnorme*, Eth. fil. Mount Esk.

Fig. 6. Do. Pinnule enlarged, showing veins and venules.

Fig. 7. *Thinnfeldia odontopteroides*, Morris, var. Portion of a pinna unsymmetrically lobed. Kilcoy Range.

Fig. 8. *Otozamites?* sp. ind. Portion of two pinnules. Burrum Coal Measures.

Fig. 9. *Glossopteris Browniana*, Brongniart. Fine-veined frond. Cooktown.

Fig. 10. *Glossopteris Browniana*, Brongniart. Large, coarsely-reticulated variety. Bowen River Coal Field. (Fresh Water or Upper Series.)

Fig. 11. *Brachyphyllum crassum*, Ten. Woods. Ipswich Coal Measures.

Fig. 12. Do. Fragment enlarged.

Fig. 13. *Phyllotheca australis*, Brongniart. Bowen-River Coal Field. (Upper or Fresh Water Series.)

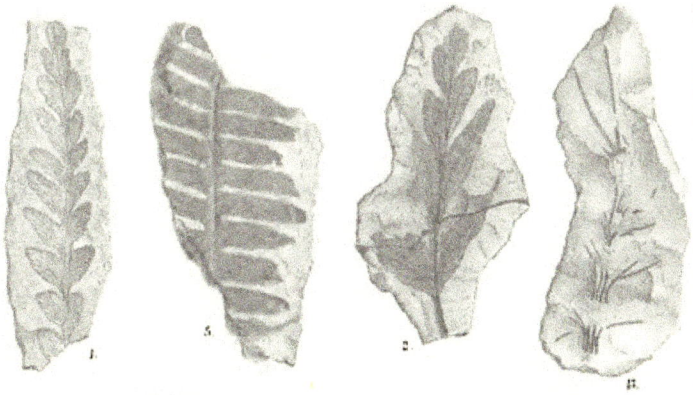

PLATE 18.

Fig. 1. *Araucarites polycarpa*, Ten. Woods? Portion of a cone. ×2. Stewart's Creek, Rockhampton.

Fig. 2. *Brachyphyllum crassum*, Ten. Woods. A small branch, but uncompressed. Ipswich Coal Measures.

Fig. 3. Do. Another specimen with the apical mucro to each leaf base. ×1½. Rosewood, near Ipswich.

Fig. 4. *Podozamites*, sp. ind. Three pinnules of a peculiar form. Burrum Coal Measures.

Fig. 5. Do. Broader pinnules than those of Fig. 4. Mount Esk.

Fig. 6. *Podozamites Kidstoni*, Eth. fil. Part of a frond of a peculiar type, with pinnules varying according to position. ×2. Burrum Coal Measures.

Fig. 7. Do. A pinnule enlarged.

Fig. 8. *Trichomanites spinifolium*, Ten. Woods. A stouter variety than the type figure. Mount Esk.

Fig. 9. *Trichomanites laxum*, Ten. Woods? Burrum Coal Measures.

Fig. 10. *Thinnfeldia media*, Ten. Woods, or *T. indica*, Feistmantel? Colinton.

Fig. 11. *Ptilophyllum oligoneurum*, Ten. Woods. Wycarbah.

Fig. 12. *Tæniopteris Daintreei*, McCoy? Portion of a frond, with direct and very slightly oblique veins. Stewart's Creek, Rockhampton.

Fig. 13. Do. Another example, with the veins less direct. Stewart's Creek, Rockhampton.

Fig. 14. *Glossopteris linearis*, McCoy. A small example. (This figure is defective.) Cooktown.

Fig. 15. *Glossopteris Browniana*, Brongniart. Portion of a frond with fine veins. Near Cooktown.

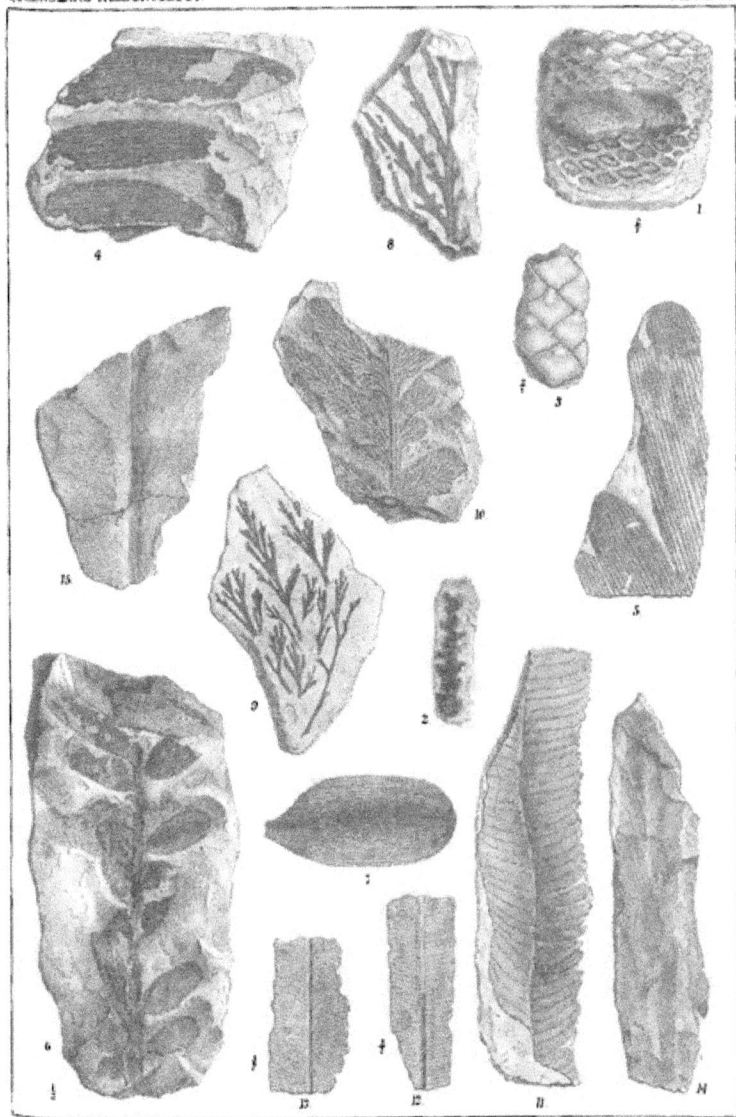

LOWER MEZOZOIC PLANTS.

PLATE 19.

Purisiphonia Clarkei, Bowerbank.

Fig. 1. A fragment of the wall of the sponge, showing the outer surface and the apertures of the canals. Natural size. Drawn from the type specimen, described by Dr. Bowerbank, and now in the British Museum of Natural History, South Kensington.

Fig. 2. The same specimen viewed laterally, showing the thickness of the wall and the course of some of the canals.

Fig. 3. A portion of the outer surface, enlarged 12 diameters, showing the disposition of the fascicles of linear spicules and the oscular apertures.

Fig. 4. A fragment of the spicular-mesh of the interior of the sponge-wall, showing its irregular character. Enlarged 60 diameters.

Figs. 5 and 6. Fragments of the fascicles of the linear spicules. Enlarged 60 diameters.

Figs. 7, 8, and 9. Entire and fragmentary hexactinellid spicules from the interspaces of the spicular-mesh. Enlarged 150 diameters.

Fig. 10. A group of minute hexactinellid spicules, confusedly intermingled and cemented together. They occur in the canals of the sponge-wall. Enlarged 60 diameters.

Fig. 11. Portions of rosette flesh-spicules. Enlarged 250 diameters. Drawn from the slide in the possession of Mr. H. J. Carter, F.R.S. [All B.M.]

G. J. H.

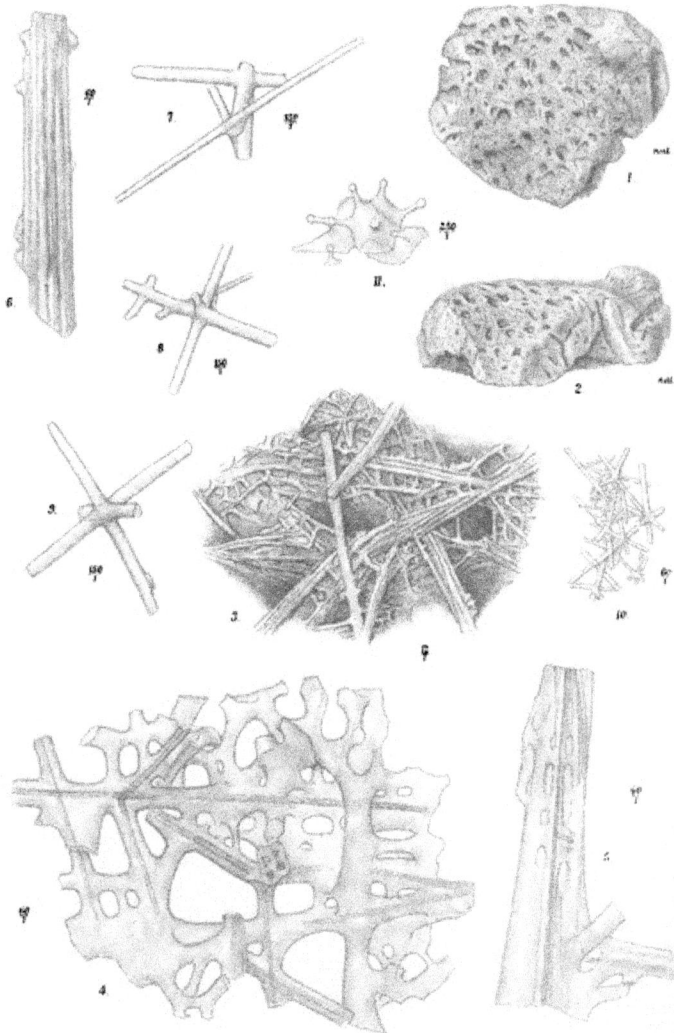

CRETACEO-JURASSIC SPONGIDA.

PLATE 20.

Fig. 1. *Pentacrinus australis*, Moore. Exterior of about two-thirds of the calyx, with portions of five arms (after Moore, *Quart. Journ. Geol. Soc.*, xxvi., t. 17, f. 3). Mitchell Downs. [W.B.C.]

Fig. 2. Do. Interior of the calyx, with portions of four arms (after Moore, *Quart. Journ. Geol. Soc.*, xxvi., t. 18, f 1). Mitchell Downs. [W.B.C.]

Fig. 3. Do. Portions of arms (after Moore, *Quart. Journ. Geol. Soc.*, xxvi., t. 18, f. 1). Mitchell Downs. [W.B.C.]

Fig. 4. *Eschara flindersensis*, H. Woodward. Proximal half of one of the posterior wings, showing nervures (after Woodward, *Geol. Mag.*, 1884, 1, t. 11, f. 1). ×2. Flinders River.

Fig. 5. *Lepralia oolitica*, Moore, encrusting a foreign body (after Moore, *Quart. Journ. Geol. Soc.*, xxvi., t. 17, f. 2). Wollumbilla. (The original is a very unsatisfactory and indefinite representation.) [W.B.C.]

Fig. 6. Do. Three cells much enlarged (after Moore, *Quart. Journ. Geol. Soc.*, xxvi., t. 17, f. 2a).

Fig. 7. *Terebratella Davidsoni*, Moore. Ventral valve (after Moore, *Quart. Journ. Geol. Soc.*, xxvi, t. 10, f. 2). Wollumbilla. [W.B.C.]

Fig. 8. Do. Dorsal valve (after Moore, *Quart. Journ. Geol. Soc.*, xxvi., t. 10, f. 1). [W.B.C.]

Fig. 9. *Argiope wollumbillaensis*, Moore. Dorsal valve (after Moore, *Quart. Journ. Geol. Soc.*, xxvi., t. 10, f. 3). ×½. Wollumbilla. [W.B.C.]

Fig. 10. Do. Ventral valve (after Moore, *Quart. Journ. Geol. Soc.*, xxvi., t. 10, f. 4). ×½. Wollumbilla. [W.B.C.]

Fig. 11. Do. Interior of the dorsal valve (after Moore, *Quart. Journ. Geol. Soc.*, xxvi., t. 10, f. 5). ×½. Wollumbilla. [W.B.C.]

Fig. 12. *Argiope punctata*, Moore. Dorsal valve (after Moore, *Quart. Journ. Geol. Soc.*, xxvi., t. 10, f. 6). Wollumbilla. Enlarged. [W.B.C.]

Fig. 13. *Rhynchonella rustica*, Moore. Dorsal valve (after Moore, *Quart. Journ. Geol. Soc.*, xxvi., t. 10, f. 7). Wollumbilla. [W.B.C.]

Fig. 14. *Lingula ovalis*, Sowerby? Half of a ventral valve (after Moore, *Quart. Journ. Geol. Soc.*, xxvi., t. 10, f. 14). Wollumbilla. [W.B.C.]

Fig. 15. *Discina apicalis*, Moore. A very unsatisfactory and indefinite figure as in the original (after Moore, *Quart. Journ. Geol. Soc.*, xxvi. t. 10, f. 15). Wollumbilla. [W.B.C.]

Fig. 16. *Pinna*, sp. ind. (compare *P. laticostata*, Stoliczka). Part of the united valves of a very large specimen. ½ nat.

Fig. 17. Do. Section of the same. ½ nat.

Fig. 18. *Maccoyella?* sp. ind. A decorticated valve which may be a species of this genus.

Fig. 19. *Rhynchonella solitaria*, Moore. Ventral valve (after Moore, *Quart. Journ. Geol. Soc.*, xxvi., t. 10, f. 10). Wollumbilla. [W.B.C.]

PLATE 20

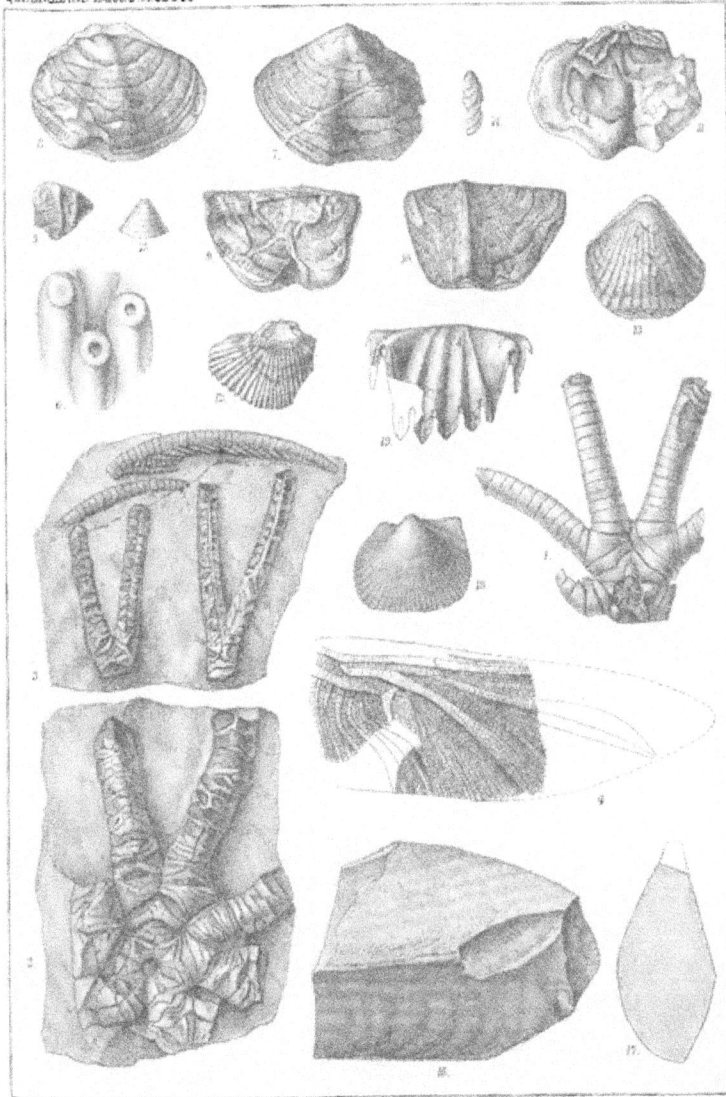

Bergass & Highley del. et lith.

McGann Bros. imp.

CRETACEO-JURASSIC ECHINODERMATA, INSECTA, & MOLLUSCA.

PLATE 21.

Fig. 1. *Ostrea vesiculosa*, J. Sowerby. Attached valve with lobe undeveloped. Chollarton.

Fig. 2. Do. Attached valve with a concentrically furrowed surface, and with remains of a lateral lobe. Chollarton.

Fig. 3. Do. Interior of Fig. 2, showing the cartilage furrows, &c. Chollarton.

Fig. 4. *Amusium*, sp. ind. (comp. *A. sulcatellum*, Stoliczka). Small portion of one valve with the ears preserved, but the general surface rather decorticated. Rockwood.

Fig. 4a. Do. Surface enlarged.

Fig. 5. *Pecten*, sp. ind. (? *P. socialis*, Moore, or *P. psila*, Ten. Woods). A small valve seen from the inside. × 2. Mitchell bore.

Fig. 6. *Pecten socialis*, Moore. Valve devoid of ornament (after Moore, *Quart. Journ. Geol. Soc.*, xxvi., t. 11. f. 9). Wollumbilla. [W.B.C.]

Fig. 7. Do. Valve with square ears and an ornament of divaricating lines, seen from the inside. Rockwood.

Fig. 8. *Pecten æquilineatus*, Moore? Fragment of a valve seen from the interior with the external reticulate structure visible through the thin shell. Aramac.

Fig. 9. *Pecten socialis*, Moore. Valve with the same general appearance as Fig. 7 and divaricating lines, but with a larger ear. × 1½. Aramac.

Fig. 10. *Pecten æquilineatus*, Moore. Valve seen from the exterior, showing surface cancellation, but ears not perfect (after Moore, *Quart. Journ. Geol. Soc.*, xxvi., t. 11, f. 11). Wollumbilla. [W.B.C.]

Fig. 11. *Pecten*, sp. ind. An ill-preserved specimen seen from the inside, thin-shelled and highly cancellated. Possibly allied to Fig. 12. Isis River.

Fig. 12. *Pecten braamburiensis*, Moore (after Moore, *Quart. Journ. Geol. Soc.*, xxvi., t. 11, f. 5). Wollumbilla. [W.B.C.]

Fig. 13. *Lima? Raudsi*, Eth. fil. A decorticated imperfect valve. Maryborough.

Fig. 14. *Undetermined Bivalve.* × 1½. Corcena Woolshed.

Fig. 15. Do. Two valves displaced, probably of the same species. × 1½. Aramac.

Fig. 16. Do. Resembling genus *Meleagrinella*, Whitfield. × 1½. Aramac.

Fig. 17. *Inoceramus Cripsi*, Mantell? Portion of a valve with concentric corrugations. Aramac Well.

Fig. 18. Do. Another specimen partially decorticated and worn.

Fig. 19. *Inoceramus*, sp. ind. (? Young of *I. pernoides*, Etheridge.) Evora.

Fig. 20. *Pecten? Moorei*, Eth. fil. Portion of a valve, probably a *Maccoyella* (*P. fimbriatus*, Moore, *Quart. Journ. Geol. Soc.*, xxvi., t. 11, f. 8). Wollumbilla. [W.B.C.]

Fig. 21. *Trigonia*, sp. ind. Interior of a partially preserved valve. Blackall.

Fig. 22. *Natica Jackii*, Eth. fil. Julia Creek.

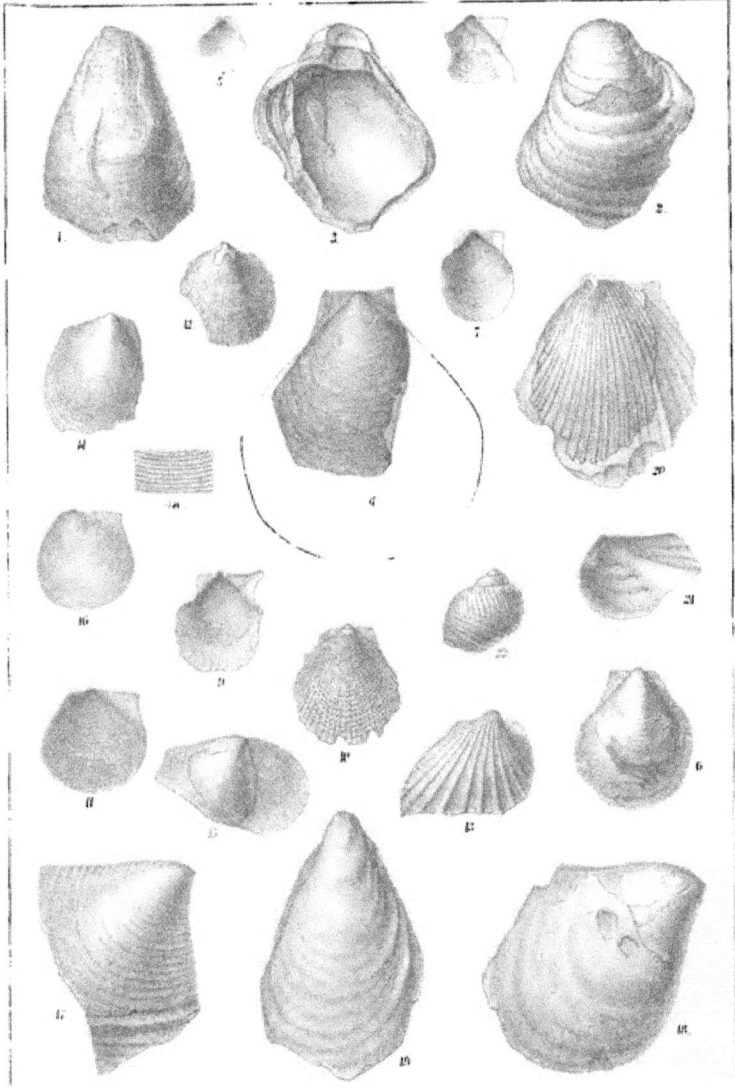

PLATE 22.

Fig. 1. *Maccoyella Barklii*, Moore, sp.? Right valve with the umbo and anterior ear corrugations of the left valve (after Moore, *Quart. Journ. Geol. Soc.*, xxvi., t. 11, f. 2). Wollumbilla. [W.B.C.]

Fig. 2. Do. Left valve with posterior flattened ear, and ridges of the anterior ear well shown (after Moore, *Quart. Journ. Geol. Soc.*, xxvi., t. 11, f. 1). Wollumbilla. [W.B.C.]

Fig. 3. Do. var. *mariæburiensis*, Eth. fil. Left valve of a well-marked variety with rugged radiating costæ, the posterior spinose. Maryborough.

Fig. 4. *Maccoyella Barklii*, Moore, sp. Interior of a right valve, showing expanded posterior wing sub-central, and supplementary muscular scars, and deeply divided anterior lobe. Maryborough.

Fig. 5. Do. A small individual, with the valves in apposition, the costæ of the right valve but faintly preserved. Walsh River.

Fig. 6. *Maccoyella umbonalis*, Moore, sp. Left valve, showing the more or less equilateral form, gibbous umbonal region, and numerous equal radiating costæ (after Moore, *Quart. Journ. Geol. Soc.*, xxvi., t. 12, f. 3). Wollumbilla. [W.B.C.]

Fig. 7. Do. Right valve (after Moore, *Quart. Journ. Geol. Soc*, xxvi., t.12, f. 2). Wollumbilla. [W.B.C.]

Fig 8. *Maccoyella corbiensis*, Moore, sp. Left valve (after Moore, *Quart. Journ. Geol. Soc.*, xxvi., t. 11, f. 7). Mount Corby. [W.B.C.]

Fig. 9. Do. Cast taken from an impression corresponding generally with this species, but possessing intermediate costæ. Maryborough.

Fig. 10. *Maccoyella? substriata*, Moore, sp. Right valve enlarged, showing the large posterior wing and intercostal ribs (after Moore, *Quart. Journ. Geol. Soc.*, xxvi., t. 11, f. 6). Wollumbilla. [W.B.C.]

PLATE 22

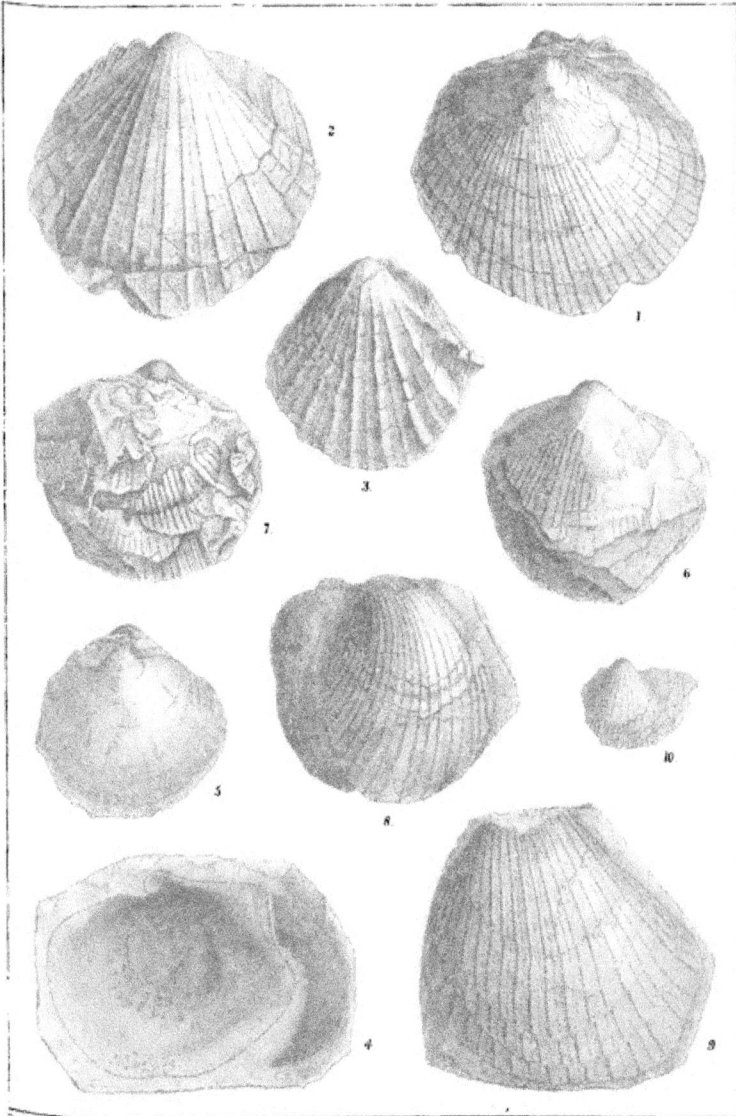

Dinjean & Ridley del et lith.

Madern Bros. imp

CRETACEO-JURASSIC PELECYPODA.

PLATE 23.

Fig. 1. *Maccoyella reflecta*, Moore, sp.? Left valve much weather-worn. Walsh River.

Fig. 2. Do. Right valve of the same specimen with the deeply divided ear.

Fig. 3. Do. Two valves mechanically united although displaced (after Moore, *Quart. Journ. Geol. Soc.*, xxvi. t. 12, f. 1). ½ nat. [W.B.C.]

Fig. 4. *Maccoyella umbonalis*, Moore, sp? Right valve of a large example, an impression in which the true shelly layers have been removed. Maryborough. It is possible that this may be a distinct species. ½ nat.

Fig. 5. *Maccoyella reflecta*, Moore, sp. Portion of a right valve, the radiating costæ obliterated by abrasion. Mount Abundance.

Fig. 6. Do. Hinge of the same specimen from the interior, with the cartilage furrows, byssal sinus, &c., visible.

Fig. 7. Do. The same, the hinge seen from above.

Fig. 8. *Maccoyella? substriata*, Moore, sp.? Probably the flat valve of Fig. 9. Hughenden.

Fig. 9. Do. An inequilateral, somewhat oblique shell, with a long upwardly curved hinge. Hughenden.

Fig. 10. *Maccoyella reflecta*, var. *Gilliatti*, Ten. Woods. Right valve, with numerous close, radiating, and somewhat thick costæ (after Woods, *Proc. Linn. Soc. N.S. Wales*, viii., t. 12, f. 5). Grey Ranges.

PLATE 23.

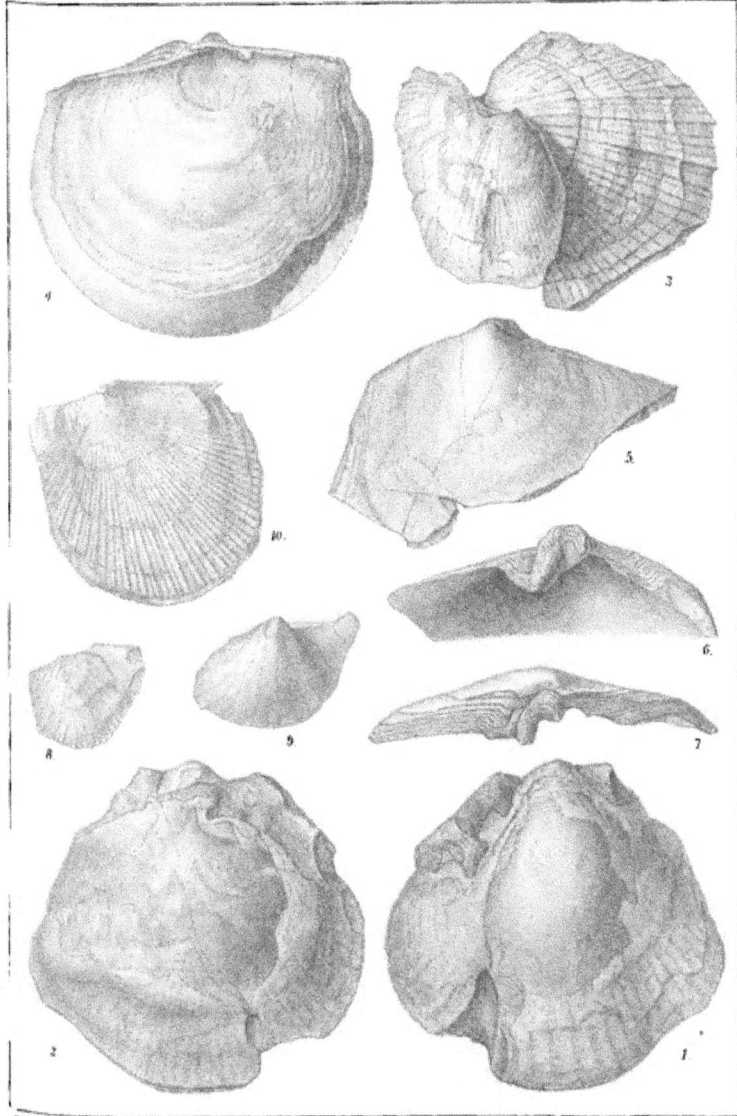

PLATE 24.

Fig. 1. *Undetermined Bivalve.* Portion of a compressed or flattened valve, without costæ, but possessing strong concentric laminæ. Richmond Downs Station.

Fig. 2. *Oxytoma? simplex*, Moore, sp. (= *Avicula æqualis*, Moore.) A correct figure of Moore's, t. 11, f. 3 (*Quart. Journ. Geol. Soc.*, xxvi.), with the posterior wing removed. ×4. Wollumbilla. [P.S.B.]

Fig. 3. Do. Left valve, much broken along the postero-ventral region, and possessing seven or eight costæ (after Moore, *Quart. Journ. Geol. Soc.*, xxvi., t. 11, f. 4). ×4. Wollumbilla. [P.S.B.]

Fig. 4. Do. Left valve with a rather smaller posterior wing, and about the same number of costæ as Fig. 3. ×2. Minmi, near Roma.

Fig. 5. Do. Left valve, with well-developed posterior wing and eight costæ (*Avicula plicata*, after Moore, *Quart. Journ. Geol. Soc.*, xxvi., t. 12, f. 6). "Enlarged." Wollumbilla. [W.B.C.]

Fig. 6. *Pseudavicula anomala*, Moore, sp. Left valve (after Moore, as *Lucina? anomala*, *Quart. Journ. Geol. Soc.*, xxvi., t. 14, f. 4). Wollumbilla. [W.B.C.]

Fig. 7. *Pseudavicula australis*, Moore, sp. Left valve (after Moore, as *Lucina australis*, *Quart. Journ. Geol. Soc.*, t. 14, f. 5). Wollumbilla. [W.B.C.]

Fig. 8. *Pseudavicula anomala*, Moore, sp. Two valves compressed one on the other. ×2. Maranoa River.

Fig. 9. *Pseudavicula australis*, Moore, sp. Left valve with worn surface, the shelly matter removed about the region of the adductor. Coottanoonna, S. Australia.

Fig. 10. Do. A small right valve.

Fig. 11. *Pseudavicula anomala*, Moore, sp. Portion of a small left valve. Maranoa River.

Fig. 12. *Pseudavicula australis*, Moore, sp. Valves in apposition, but rather displaced, showing hinge. ×2. Coottanoonna, S. Australia.

Fig. 13. Do. Right valve much worn (after Hudleston, as *Avicula orbicularis*, *Geol. Mag.*, 1884, 1, t. 11, f. 10). Peak District, South Australia. [B.M.]

Fig. 14. *Pseudavicula alata*, Etheridge. Decorticated left valve (after Etheridge, as *Avicula alata*, *Quart. Journ. Geol. Soc.*, xxviii., t. 20, f. 8). Maryborough. [R.D.]

Fig. 15. *Oxytoma rockwoodensis*, Eth. fil. Left valve. Rockwood.

Fig. 16. *Lima Gordoni*, Moore. Left valve (after Moore, *Quart. Journ. Geol. Soc.*, xxvi., t. 12, f. 4). Wollumbilla. [W.B.C.]

Fig. 17. *Lima? multistriata*, Moore (after Moore, *Quart. Journ. Geol. Soc.*, xxvi., t. 12, f. 5). Wollumbilla. [W.B.C.]

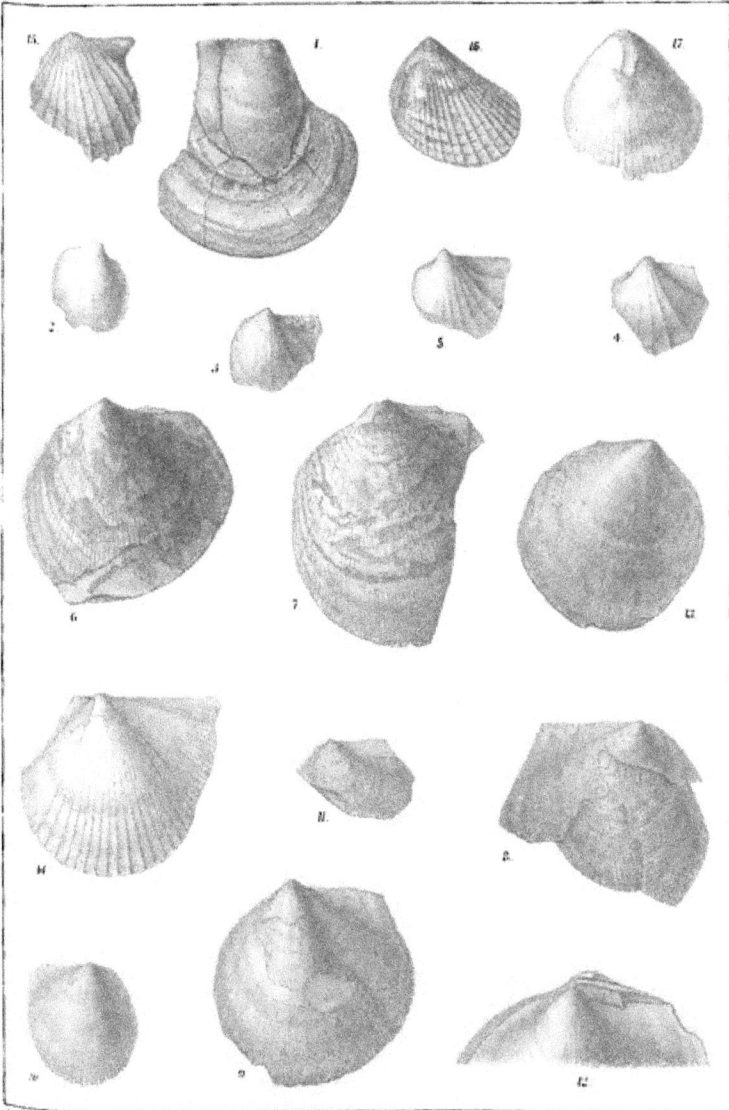

PLATE 25.

Fig. 1. *Aucella hughendenensis*, Etheridge. Left valve (after Etheridge, *Quart. Journ. Geol. Soc.*, xxviii., t. 25, f. 3). Hughenden. [R.D.]

Fig. 2. Do. Surface sculpture of the same. ×4.

Fig. 3. Do. Left valve somewhat imperfect. Barcoo River.

Fig. 4. Do. Right valve, with the anterior ear broken off. Barcoo River.

Fig. 5. Do. A similar, but smaller and rather more transverse example. ×2. Upper Flinders River.

Fig. 6. Do. Right valve, with the anterior ear preserved. Blackall Well.

Fig. 7. *Inoceramus pernoides*, Etheridge. Mount Cornish Homestead.

Fig. 8. Do. With coarse concentric rugæ. Upper Flinders River.

Fig. 9. *Inoceramus Carsoni*, McCoy. ½ nat. Upper Flinders River.

Fig. 10. Do. Upper Flinders River.

Fig. 11. *Mytilus inflatus*, Moore. Right valve (after Moore, *Quart. Journ. Geol. Soc.*, xxvi., t. 13, f. 4). Wollumbilla. [W.B.C.]

Fig. 12. *Inoceramus pernoides*, Etheridge. Left valve (after Etheridge, *Quart. Journ. Geol. Soc.*, xxviii., t. 22, f. 23). Marathon. [R.D.]

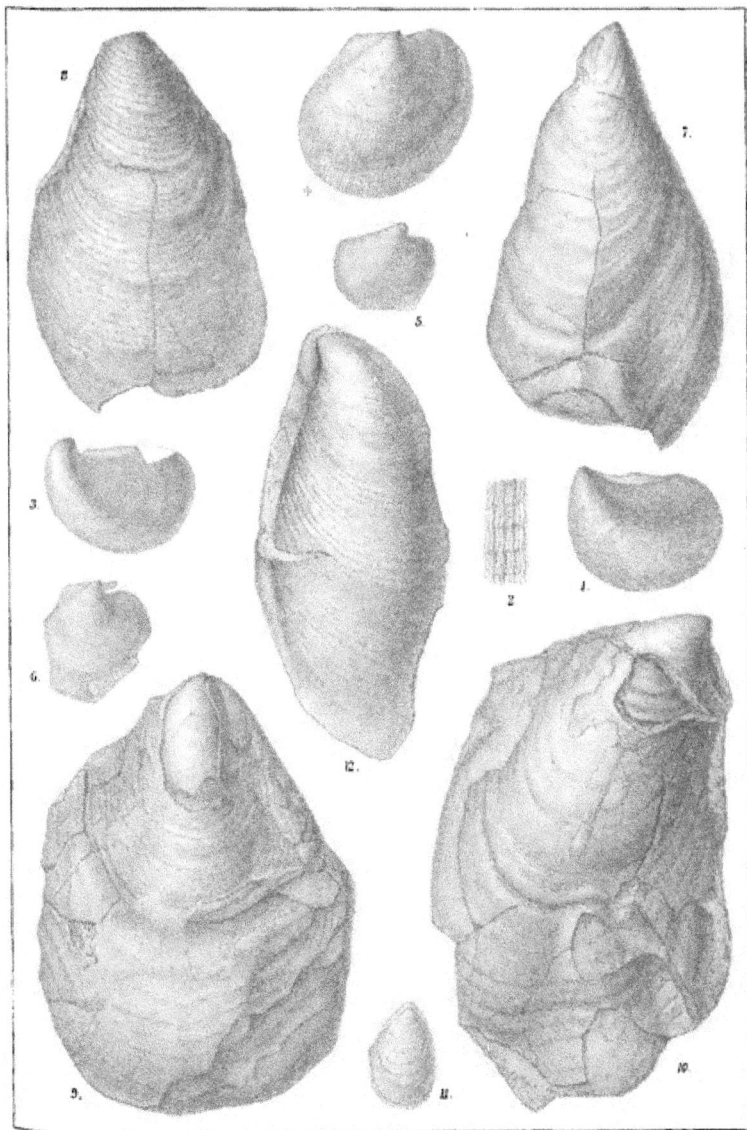

CRETACEO-JURASSIC PELECYPODA.

PLATE 26.

Fig. 1. *Cucullæa robusta*, Etheridge. Artificial cast of the right valve, taken from an impression. Maryborough.

Fig. 2. *Cucullæa Hendersoni*, Eth. fil. Internal cast, with hinge teeth exposed. Glanmire Block, near Tambo.

Fig. 3. Do. Right valve partially preserved. Glanmire Block, near Tambo.

Fig. 4. *Cucullæa robusta*, Etheridge. Internal cast of a small or young left valve. Maryborough.

Fig. 5. *Trigonia*, sp. ind. Internal cast of a left valve. Maryborough.

Fig. 6. *Nucula gigantea*, Etheridge. Internal cast of a right valve, with hinge teeth. Port Curtis.

Fig. 7. Do. Another right valve. Maryborough.

Fig. 8. *Nucula quadrata*, Etheridge. Internal cast of left valve. Aramac.

Fig. 9. Do. View of hinge, showing casts of the deep anterior muscular impressions. Aramac.

Fig. 10. *Nuculana (Yoldia?) Randsi*, Eth. fil. Internal cast of right valve. Isis River.

Fig. 11. *Cytherea?* sp. ind. Much decorticated and ill-preserved left valve. Erora Station. Compare *Cytherea? Hudlestoni* (t. 28, f. 12).

Fig. 12. Do. View of hinge of the same.

Fig. 13. *Undetermined Genus* (? *Unicardium*). Left valve, with the test preserved. Upper Flinders River.

Fig. 14. Do. Interior of the left valve of another individual. Upper Flinders River.

Fig. 15. Do. View of the hinge of Fig. 13.

Fig. 16. *Palæomæra*, sp. ind. A rather compressed and partially decorticated example. Muttaburra.

Fig. 17. Do. A cast. Maryborough.

Fig. 18. *Cyprina Clarkei*, Moore? Left valve of a poor cast. ½ nat. Maryborough.

Fig. 19. Do. View of hinge of the same specimen.

Fig. 20. *Ceromya?* sp. ind. Right valve of a poor cast, rather crushed. ½ nat. Maryborough.

Burjass & Highley del. et lith.

Mintern Bros. imp.

CRETACEO-JURASSIC PELECYPODA.

PLATE 27.

Fig. 1. *Unicardium? Etheridgei*, Eth. fil. Maryborough. ½ nat.

Fig. 2. *Mactra Mecki*, Eth. fil. Left valve, with the test partially preserved. Walsh River.

Fig. 3. Do. Hinge of another specimen. Walsh River.

Fig. 4. *Corbicella? maranoana*, Eth. fil. An almost perfect left valve. Maranoa River.

Fig. 5. Do. Interior of hinge.

Fig. 6. *Macrocallista plana*, Moore, sp. Internal cast of right valve. Walsh River.

Fig. 7. Do. Hinge. Walsh River.

Fig. 8. Do. Internal cast of the left valve of a rather shorter variety. Minmi, near Roma.

Fig. 9. *Cyprina Clarkei*, Moore. Internal cast of the left valve. 1½ nat. Walsh River.

Fig. 10. *Cyprina Clarkei*, Moore? Internal cast of the right valve, with pallial impression and sinus. Walsh River.

Fig. 11. Do. Internal cast of another example. ½ nat. Walsh River.

Fig. 12. *Cytherea Woodwardiana*, Hudleston. Right valve, with test partially preserved (after Hudleston, *Geol. Mag.*, 1884, i., t. 11, f. 8a). [B.M.]

Fig. 13. Do. View of anterior end of the same specimen (after Hudleston, *Geol. Mag.*, 1884, i., t. 11, f. 8b). [B.M.]

Fig. 14. Do. Cast of the interior of another example, with the scars of the adductors, pallial line, and sinus visible (after Hudleston, *Geol. Mag.*, 1884, i., t. 11, f. 8c). [B.M.]

Fig. 15. *Astarte apicalis*, Moore. Right valve, with the test apparently preserved (after Moore, *Quart. Journ. Geol. Soc.*, xxvi., t. 13, f. 11). Western Australia. [W.B.C.]

Fig. 16. *Astarte wollumbillensis*, Moore. Right (?) valve, very imperfect (after Moore, *Quart. Journ. Geol. Soc.*, xxvi., t. 12, f. 12). Wollumbilla. [W.B.C.]

Fig. 17. *Mactra trigonalis*, Moore. Right valve (after Moore, *Quart. Journ. Geol. Soc.*, xxvi., t. 14, f. 6). [W.B.C.]

Fig. 18. *Glycimeris sulcata*, Etheridge, var. Cast of left valve. Maryborough.

Benjoan & Rigney del et lith.   Mackern Bros. imp.

CRETACEO-JURASSIC PELECYPODA.

PLATE 28.

Fig. 1. *Unio eyrensis*, Tate? Portion of the left valve, with the umbone decorticated. Bundanba.

Fig. 2. *Corbicella? maranoana*, Eth. fil. Internal cast of the left valve. Near Mitchell Railway Station.

Fig. 3. Do. Interior of hinge of the same.

Fig. 4. *Glycimeris rugosa*, Moore, sp. Portion of left valve, a cast. Maranoa River.

Fig. 5. Do. Hinge and united valves of same specimen.

Fig. 6. *Glycimeris*, sp. ind. Right valve, probably of a separate species. Minmi, near Roma.

Fig. 7. *Glycimeris aramacensis*, Eth. fil. Greater portion of left valve. Aramac.

Fig. 8. Do. Hinge and united valves of the same.

Fig. 9. *Goniomya depressa*, Moore. Portion of a right (?) valve (after Moore, *Quart. Journ. Geol. Soc.*, xxvi., t. 13, f. 6). Wollumbilla. [W.B.C.]

Fig. 10. *Corimya Wilsoni*, Moore, sp. Nearly perfect left valve. (The specimen badly figured by Moore, *Quart. Journ. Geol. Soc.*, xxvi., t. 14, f. 8). Amby River. [P.S.B.]

Fig. 11. *Corimya primula*, Hudleston? Left valve of a smaller and probably distinct species. Minmi, near Roma.

Fig. 12. *Cytherea? Hudlestoni*, Eth. fil. Right valve, with the test partially preserved. Glammire Block, near Tambo.

Fig. 13. *Rocellaria terra-reginæ*, Eth. fil. A group. × 1½. Burrum Coal Field.

Fig. 14. Do. A single individual. × 1½.

PLATE 28

PLATE 29.

Fig. 1. *Ammonites Daintreei*, Etheridge. View of the aperture, &c. (after Etheridge, *Quart. Journ. Geol. Soc.*, xxviii, t. 24, f. 1, *bis*). ⅔ nat. Hughenden. [R.D.]

Fig. 2. Do. Side view of the same (after Etheridge, *Quart. Journ. Geol. Soc.*, xxviii., t. 24, f. 1). ⅔ nat. [R.D.]

Fig. 3. Do. Another specimen (after Etheridge, *Quart. Journ. Geol. Soc.*, xxviii., t. 24, f. 2). [R.D.]

Fig. 4. *Ammonites Sutherlandi*, Etheridge. Side view (after Etheridge, *Quart. Journ. Geol. Soc.*, xxviii., t. 21, f. 4). Marathon. [R.D.]

Fig. 5. *Ammonites olene*, Ten. Woods. View of the mouth, back, and keel (after Woods, *Journ. R. Soc. N.S. Wales*, 1882, xvi., t. 7, f. 2). Palmer River.

Fig. 6. *Belemnites Canhami*, Tate? Apex of the guard. Aramac.

Fig. 7. Do. Cross section of the same example.

Fig. 8. *Cinulia Hochstetteri*, Moore (after Moore, *Quart. Journ. Geol. Soc.*, xxvi., t. 10, f. 19). "Rather enlarged." Wollumbilla. [W.B.C.]

Fig. 9. *Actæon depressus*, Moore (after Moore, *Quart. Journ. Geol. Soc.*, xxvi., t. 10, f. 20). "Rather enlarged." Wollumbilla. [W.B.C.]

Fig. 10. *Pleurotomaria Cliftoni*, Etheridge (after Etheridge, *Quart. Journ. Geol. Soc.*, xxviii., t. 25, f. 4). Gordon Downs. [R.D.]

Fig. 11. *Natica ornatissima*, Moore (after Moore, *Quart. Journ. Geol. Soc.*, xxvi., t. 10, f. 16). Wollumbilla. [W.B.C.]

Fig. 12. *Delphinula reflecta*, Moore (after Moore, *Quart. Journ. Geol. Soc.*, xxvi., t. 10, f. 21). "Enlarged." Wollumbilla. [W.B.C.]

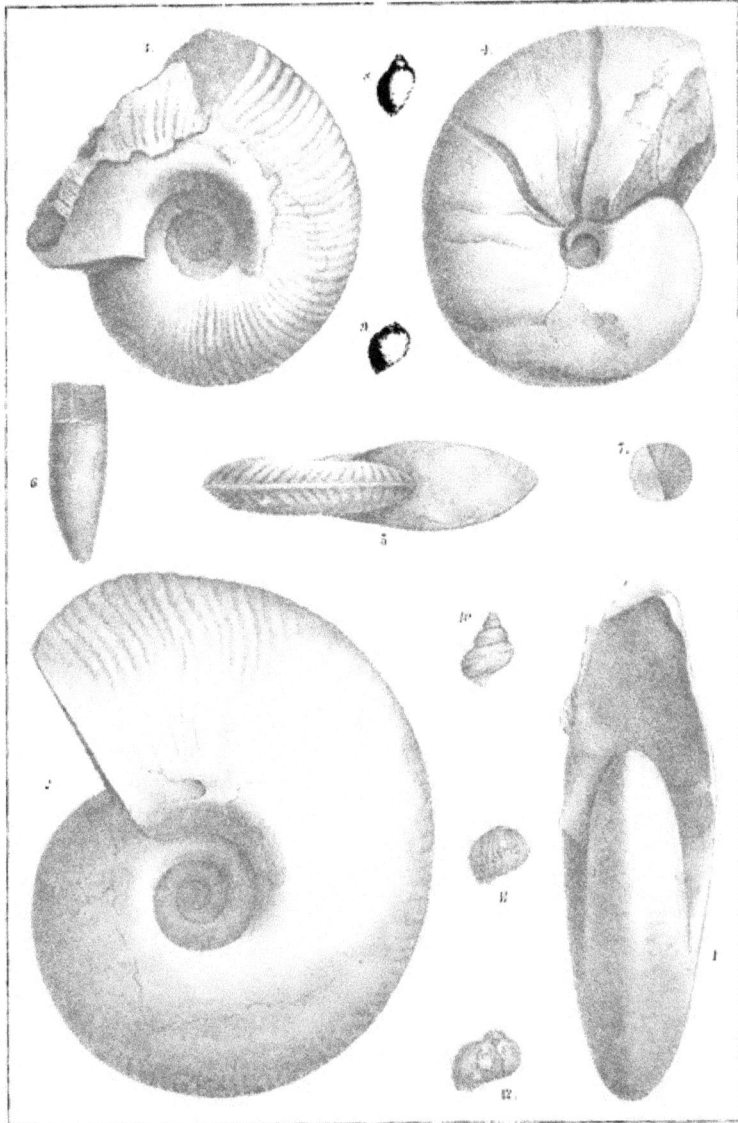

CRETACEO-JURASSIC CEPHALOPODA & GASTEROPODA.

PLATE 30.

Fig. 1. *Ammonites Flindersi*, McCoy (*A. Bendanti*, var. *Mitchelli*, Etheridge, after Etheridge, *Quart. Journ. Geol. Soc.*, xxviii., t. 23, f. 1). ½ nat. Hughenden. [R.D.]

Fig. 2. Do. (After Etheridge, *Quart. Journ. Geol. Soc.*, xxviii., t. 23, f. 1 *bis*). ¾ nat. [R.D.]

Fig. 3. Do. (After Etheridge, *Quart. Journ. Geol. Soc.*, xxviii., t. 23, f. 2). [R.D.]

Fig. 4. *Ammonites olene*, Ten. Woods. (After Woods, *Journ. R. Soc. N.S. Wales*, 1882, xvi., t. 8, f. 1.)

Fig. 5. *Ammonites*, sp. ind. (or young form of *A. Flindersi*, McCoy?) ×2. Aramac.

Fig. 6. Do. Back, with keel.

Fig. 7. *Crioceras australe*, Moore. The inner whorls (really the oldest) with tubercles. Walsh River.

Fig. 8. *Crioceras Edkinsi*, Eth. fil. Mount Cornish Station.

Fig. 9. Do. Back of the same specimen.

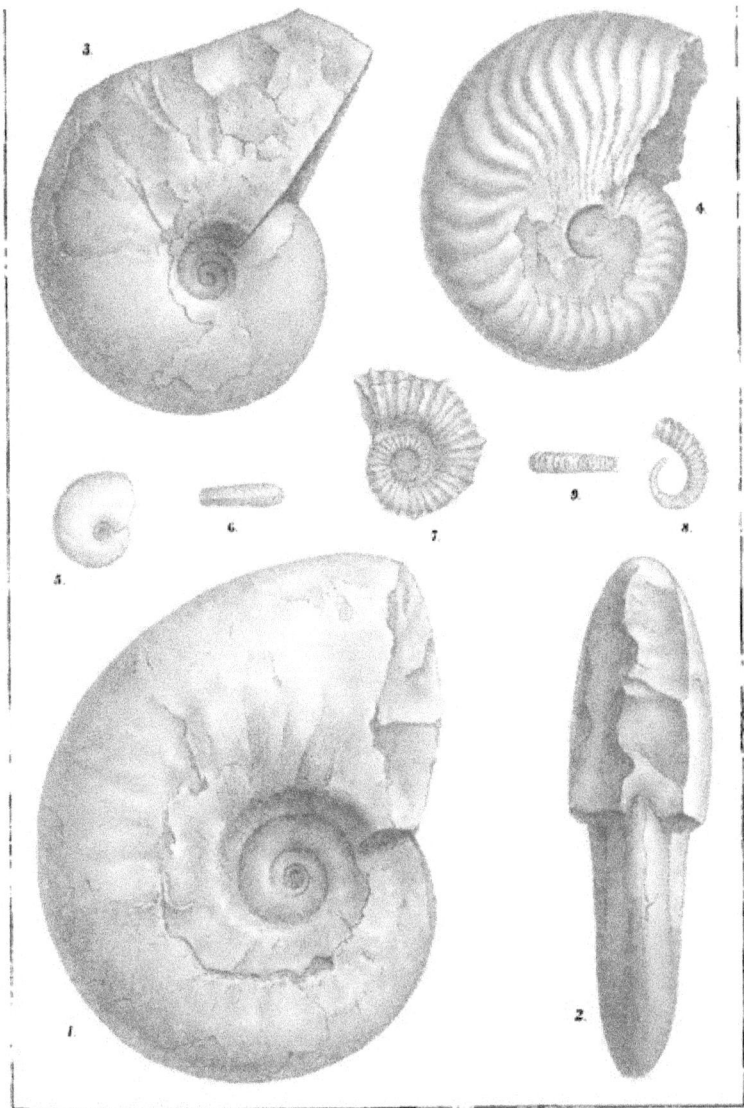

PLATE 31.

Fig. 1. *Crioceras australe*, Moore. The largest specimen but three known to the Author. Walsh River.

Fig. 2. *Natica variabilis*, Moore (*N. lineata*, after Etheridge, *Quart. Journ. Geol. Soc.*, xxviii., t. 21, f. 1). Maryborough. [R.D.]

Fig. 3. Do. (After Moore, *Quart. Journ. Geol. Soc.*, xxvi., t. 10, f. 15.) Wollumbilla. [W.B.C.]

Fig. 4. *Anchura? Wilkinsoni*, Eth. fil. Front view. Evora Station.

Fig. 5. Do. The same specimen, back view.

Fig. 6. *Gastrochæna australis*, Eth. fil. Fragment of the tube. Marauoa River.

Fig. 7. Do. Cross section of Fig. 6.

Fig. 8. Do. Surface of the tube much enlarged, showing anastomosing lines, caused by the wearing down of the shelly-frills.

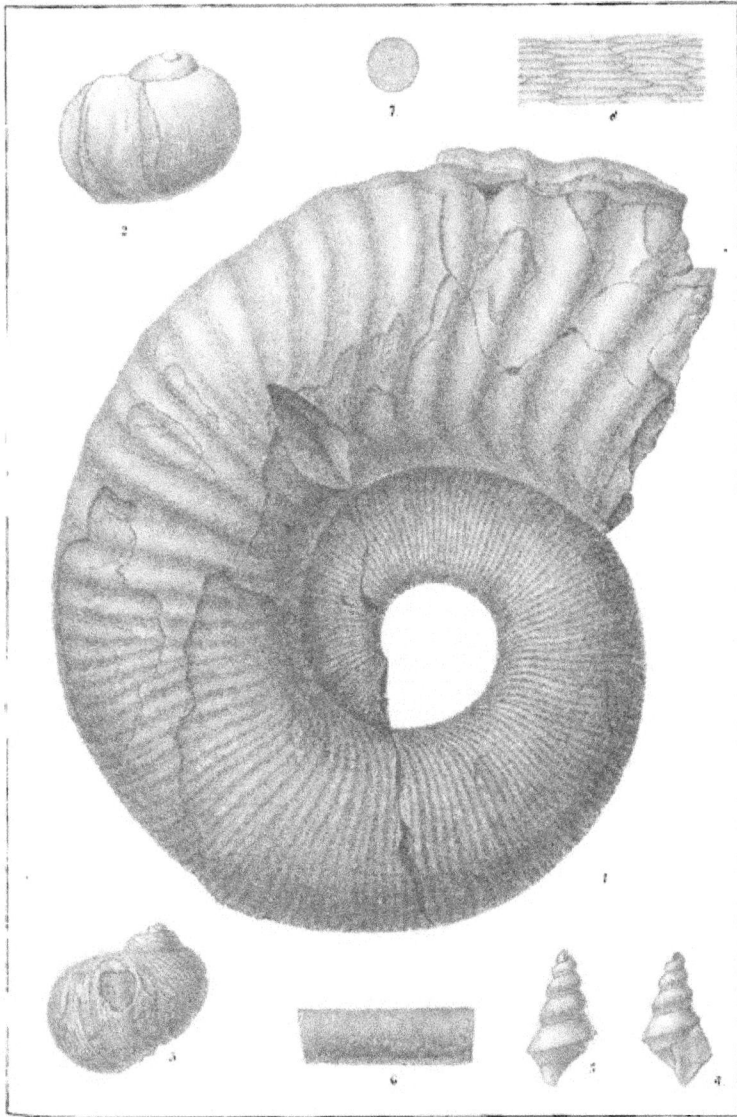

CRETACEO-JURASSIC CEPHALOPODA, GASTEROPODA & PELECYPODA.

Fig. 1. *Crioceras australe*, Moore. Side view of the older whorls, with characteristic costation and tubercles. (This and the following figures are the types of *Crioceras Jackii*, Eth. fil.) Walsh River.

Fig. 2. Do. Back of the same specimen, with the double row of tubercles. Walsh River.

Fig. 3. Do. Front view of a portion of another example, showing the flattened dorsal side, ventral siphuncle, and the lobes and saddles of one of the septa. ½ nat. Walsh River.

Fig. 4. Do. Section of the lobes and saddles taken from a portion of Fig. 3.

Fig. 5. Do. Fragment of a large individual, with a more open coil, and having the appearance of Woods' *C. irregulare*. Walsh River.

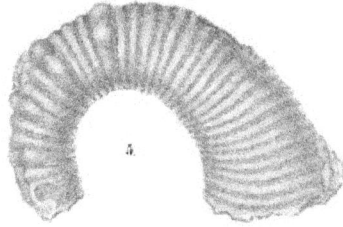

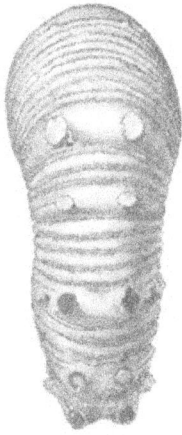

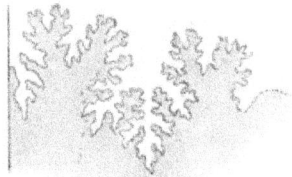

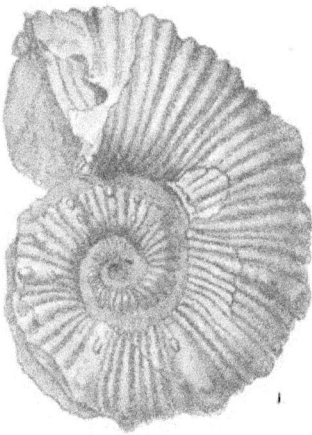

PLATE 33.

Fig. 1. *Crioceras irregulare*, Ten. Woods (? = *C. australe*, Moore). An open and uncoiled form, with costæ and tubercles like those of *C. australe* (after Ten. Woods, *Journ. R. Soc. N. S. Wales*, 1882, xvi., t. 8, f. 2). Palmer River.

Fig. 2. *Crioceras australe*, Moore. (After Moore, *Quart. Journ. Geol. Soc.*, xxvi., t. 15, f. 3). Upper Maranoa. [W.B.C.]

Fig. 3. *Ancyloceras Flindersi*, McCoy. Portion representing one of the curved parts of the shell. ½ nat. Landsborough Creek, accompanied by *Aucella hughendenensis*, Eth.

Fig. 4. *Crioceras*, sp. ind. Portion of an individual, with the costæ in bundles of two and three springing from a row of tubercles on the inner edges of the whorls. Landsborough Creek.

Fig. 5. *Crioceras*, sp. ind. A fragment with widely separated ribs and two rows of tubercles on the back. Aramac.

Fig. 6. Do. Side view of the same fragment, the costæ united and interrupted in their course by a simple tubercle.

Fig. 7. *Arca prolonga*, Moore. (After Moore, *Quart. Journ. Geol. Soc.*, xxvi., t. 14, f. 7.) [W.B.C.]

Fig. 8. *Adrana? elongata*, Etheridge, sp. (After Etheridge, *Quart. Journ. Geol. Soc.*, xxviii., t. 20, f. 5.) Maryborough. [R.D.]

Fig. 9. *Nucula truncata*, Moore. (After Moore, *Quart. Journ. Geol. Soc.*, xxvi., t. 12, f. 9.) [W.B.C.]

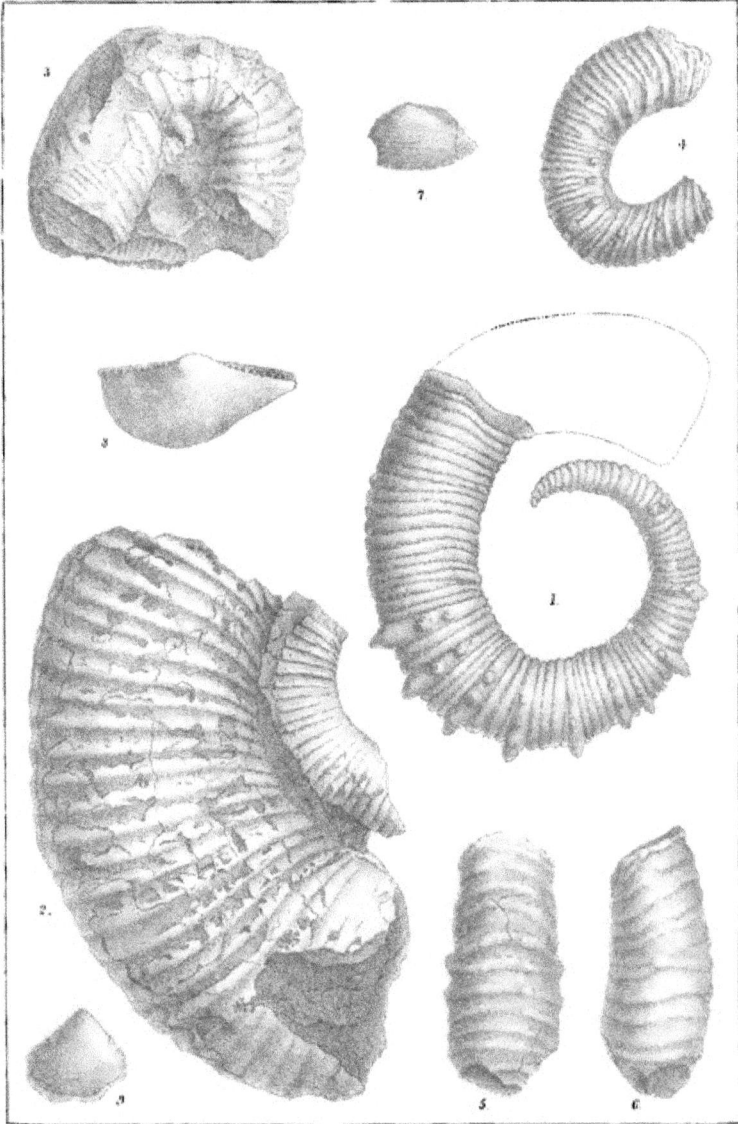

CRETACEO-JURASSIC CEPHALOPODA & PELECYPODA.

PLATE 31.

Fig. 1. *Ammonites (Schloenbachia) inflatus*, J. Sowerby? Lateral view of a well-preserved example. ½ nat. Glanmire Block, near Tambo.

Fig. 2. Do. Portion of the back, showing keel. ½ nat.

Fig. 3. Do. A septum, with lobes and saddles from Fig. 1.

Fig. 4. Do.? Artificial cast from an impression differing but little from this species. Aramac.

Fig. 5. *Ancyloceras Flindersi*, McCoy? Portion of a shell. ½ nat. Aramac Well.

Fig. 6. Do. View of the back or ventral aspect of the same specimen. ½ nat.

Fig. 7. Do. Septum seen in section and outline of the whorl. ½ nat.

Fig. 8. Do. Septum, with lobes and saddles. ½ nat.

Fig. 9. *Corbicula burrumensis*, Eth. fil. Right valve. Burrum Coal Field.

Fig. 10. Do. Hinge and united valves of the same specimen.

Fig. 11. *Nucula Cooperi*, Moore. (After Moore, *Quart. Journ. Geol. Soc.*, xxvi., t. 12, f. 5.) Wollumbilla. [W.B.C.]

Fig. 12. *Cytherea (Cyprina?) Moorei*, Eth. fil. Left valve. Mount Hamilton, South Australia.

Fig. 13. Do. Anterior view of the united valves.

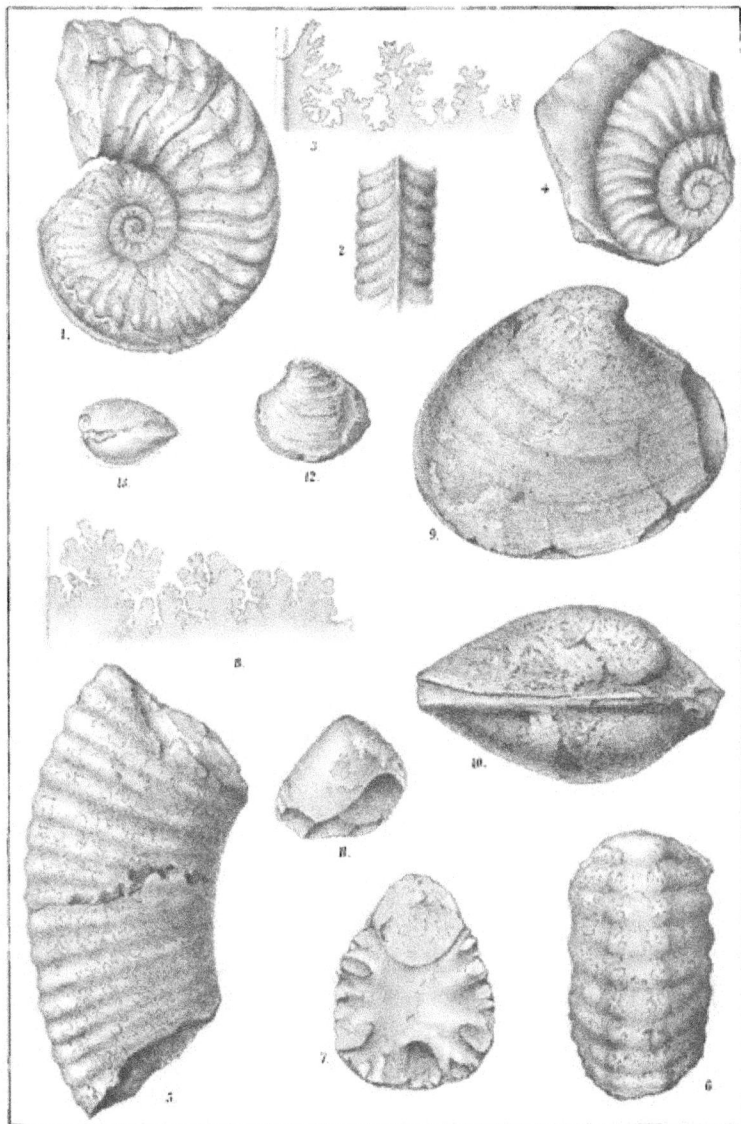

PLATE 35.

Fig. 1. *Belemnites australis*, Phillips.   Internal view of the guard (after Moore, *Quart. Journ. Geol. Soc.*, xxvi., t. 16, f. 1).   Wollumbilla. [W.B.C.]

Fig. 2.   Do.   Ventral aspect of the guard, with the two lateral grooves visible at the proximal end (after Moore, *Quart. Journ. Geol. Soc.*, xxvi.. t. 16, f. 2).   Wollumbilla.   [W.B.C.]

Fig. 3. *Belemnites Cankami*, Tate.   Ventral aspect, with the lateral grooves visible at the proximal end (*B. australis*, after Moore, *Quart. Journ. Geol. Soc.*, xxvi., t. 16, f. 3).   Ward Creek.   [W.B.C.]

Fig. 4.   Do.   Lateral view of same specimen, with one of the curved lateral grooves (after Moore, *Quart. Journ. Geol. Soc.*, xxvi., t. 16, f. 4).   [W.B.C.]

Fig. 5.   Do.   Cross section of the guard behind the alveolar apex (after Moore, *Quart. Journ. Geol. Soc.*, xxvi., t. 16, f. 5).   [W.B.C.]

Fig. 6. *Belemnites ereuus*, Tate.   Transverse section of the guard, showing alveolar chamber and siphuncle (*B. paxillosus*, after Moore, *Quart. Journ. Geol. Soc.*, xxvi., t. 16, f. 6a).   Wollumbilla. [W.B.C.]

Fig. 7. *Belemnites Cankami*, Tate ?   Dorsal view of a somewhat mucronate variety, fractured across the guard, at about the flexure of the lateral grooves.   Cambridge Downs Run.

Fig. 8.   Do.   Lateral view of the same specimen, showing the somewhat compressed form, the lateral grooving having become obliterated.

Fig. 8a.   Do.   Cross section of the guard, almost circular, at the broken proximal end.

Fig. 9.   Do.   Dorsal aspect of a form more approaching the type, with both furrows visible at the proximal end of the guard.   Aramac Well.

Fig. 10. *Belemnites Sellheimi*, Ten. Woods.   Concave surface of the first of the five septa represented in Fig. 11, showing the position of the siphuncle.   Flinders River, near Hughenden.

Fig. 11.   Do   Five chambers of the phragmacone, seen partially from the dorsal side.

Fig. 12. *Belemnites Cankami*, Tate.   Dorsal aspect with the lateral furrows visible (after Tate, *Trans. R. Soc. S.*, 1880, iii., t. 4, f. 2b).   Peak Creek, Central Australia.   [A.M.]

Fig. 13.   Do.   Lateral view of the same, showing one of the lateral furrows with its flexure and bifurcation (after Tate, *Trans. R. Soc. S. Austr.*, 1880, iii., t. 4, f. 2a.   [A.M.]

Fig. 14.   Do.   Section of the alveolus, showing the unsymmetrical portion marked off by the lateral furrows, thin test on the ventral side, and the small siphuncle (after Tate, *Trans. R. Soc. S. Austr.*, 1880, iii., t. 4, f. 2c.   [A.M.]

Fig. 15. *Belemnites*, sp. ind.   Phragmacone of about sixteen chambers in the alveolar cavity of the guard.   Blackall Road.

Fig. 16.   Do.   Another phragmacone in a similar position, but with closer and more numerous chambers.   Blackall Road.

Fig. 17. *Belemnites ? Liversidgei*, Eth. fil.   Fusiform guard with the axis visible through the shelly matter as a thin white line.   Aramac.

Fig. 18. *Belemnites*, sp. ind.   Portion of the guard with the phragmacone in the alveolar cavity, showing a fine grooved line, extending through the shelly layers.   × 2.   Aramac.

Fig. 19. *Belemnites Liversidgei*, Eth. fil.   Portion of a guard traversed by canal.   × 2.   Aramac.

Fig. 20.   Do.   A guard with an extended mucro-like distal end.   × 2.   Aramac.

Fig. 21. *Teuthis ?* sp. ind.   Portions of the shaft of a pen (after Moore, *Quart. Journ. Geol. Soc.*, xxvi., t. 16, f. 8).   [W.B.C.]

PLATE 80

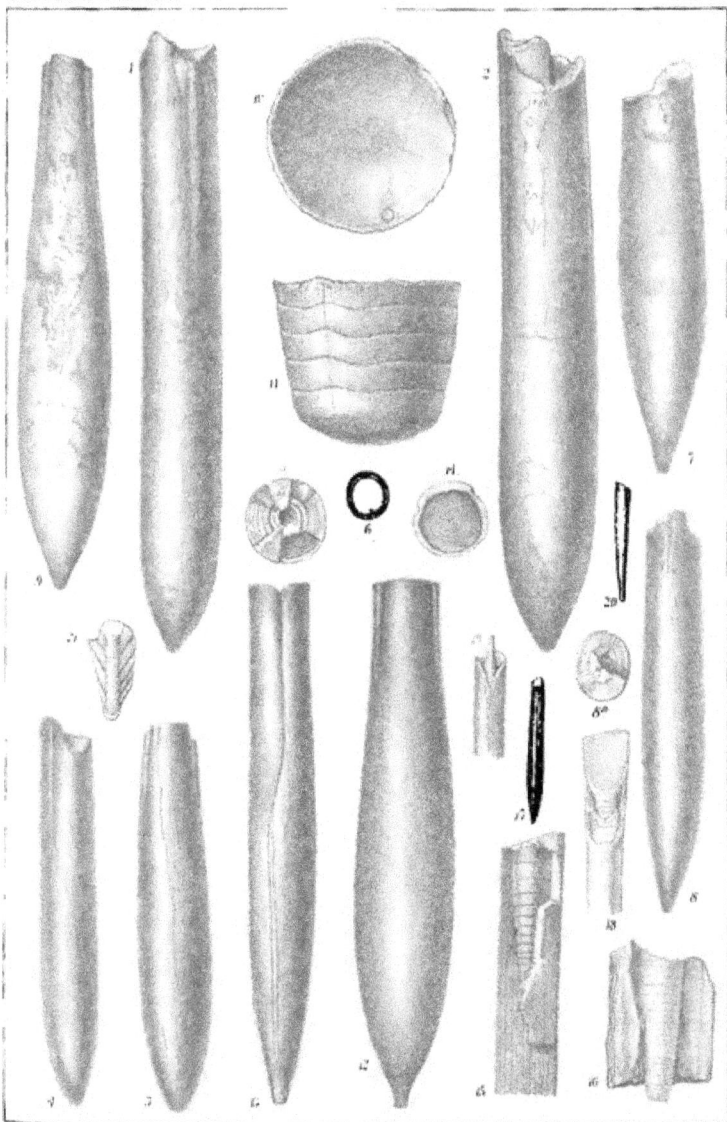

CRETACEO-JURASSIC CEPHALOPODA.

PLATE 36.

Fig. 1. *Alecopora*, sp. ind. Natural longitudinal section of part of a corallite, with the trabecular septa and spurious columella. × 3. New Guinea.

Fig. 2. Do. Horizontal view of a corallite. × 3.

Fig. 3. *Galaxea*, sp. ind. (near *G. clarus*). Natural section, showing corallites and perithcea. × ½. New Guinea.

Fig. 4. *Deltocyathus*, sp. ind. Part of a corallum. × ½. New Guinea.

Fig. 5. *Leptoria*, sp. ind. Portion of a corallum, seen horizontally. × ½. New Guinea.

Fig. 6. *Thalassina Emerii*, Bell? Abdominal segments and anterior appendages, preserved in a nodule. Cleveland Bay.

Fig. 7. *Ostrea pes-tigris*, Hanley. Upper valve, seen from the inside. Child's Vineyard.

Fig. 8. Do. Another upper valve, seen from the exterior. Child's Vineyard.

Fig. 9. Do. Lower or flat valve, from the inside.

Fig. 10. *Anomalocardia trapezia*, Deshayes. Exterior view of a right valve. Child's Vineyard.

Fig. 11. Do. Interior of a right valve. Child's Vineyard.

Fig. 12. Do. Right valve of a more rotund variety. Child's Vineyard.

Fig. 13. *Natica plumbea*, Lamarck. Child's Vineyard.

Fig. 14. *Potamides ebininus*, Bruguière. Back view. Child's Vineyard.

Fig. 15. Do. Front view. Child's Vineyard.

Fig. 16. Do. Side view of the mouth, with a much thickened outer lip. Child's Vineyard.

PLATE 36

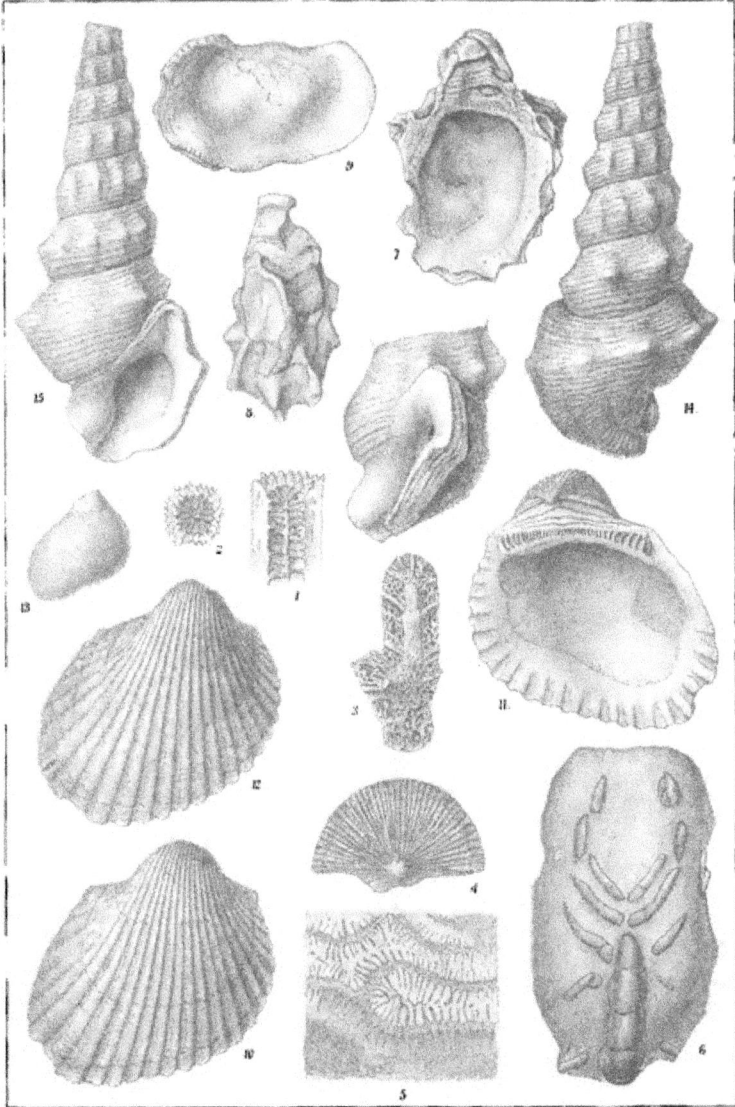

Bergasu & Highley del et lith.   Michaels Bros imp.

TERTIARY CORALS & POST TERTIARY MOLLUSCA, &c.

PLATE 37.

Fig. 1. *Striatopora ? uniseptata*, Eth. fil. Portion of a bifurcating branch. × 2. Fanning River.

Fig. 2. Do. A calice, highly magnified, showing the single septum.

Fig. 3. *Spirifera curvata*, Schlotheim. Ventral valve. Fanning Old Station.

Fig. 4. Do. Dorsal valve, and umbo of the ventral. Fanning Old Station.

Fig. 5. Do. Side view of the united valves.

Fig. 6. *Pentamerus brevirostris*, Phillips. Section of the ventral valve, exposing the septum. Fanning Old Station.

Fig. 7. *Spirifera curvata*, Schlotheim. Portion of a decorticated ventral valve. Fanning River.

Fig. 8. *Atrypa desquamata*, Sby. A much worn and rather distorted ventral valve. Broken River.

Fig. 9. *Pentamerus brevirostris*, Phillips. A variety, with indistinct costæ on the flattened and undefined fold. Fanning Old Station.

Fig. 10. Do. Side view of united valves.

Fig. 11. Do. Dorsal valve and umbo of the ventral of another specimen, showing a strongly-produced sulcus in front, slit of the septum on the ventral umbo, and those of the dividing lamellæ on the dorsal. Fanning Old Station.

Fig. 12. *Orthotetes concentrica*, Eth. fil. Ventral valve, with a high umbo and concentric laminæ. Fanning Old Station.

Fig. 13. Do. Dorsal valve of the same, with very fine radiating striæ. Fanning Old Station.

Fig. 14. Do. Side view of the united valves.

Fig. 15. *Amplexus*, sp. ind. Side view of calice. Broken River.

Fig. 16. Do. Horizontal section, with the marginal septa, and large central tabulate area. × 2.

Fig. 17. *Spirifera bicarinata*, Eth. fil. Ventral valve and one of the alar expansions. × 1½. Keelbottom.

Fig. 18. *Productus*, sp. ind. (resembling *P. Humboldti*, D'Orb). Ventral valve, with coarse elongated spines. Mount Britton Gold Field.

Fig. 19. *Chonetes*, sp. ind. Internal cast of the dorsal valve. Mount Britton Gold Field.

Fig. 20. *Chonetes*, sp. ind. Impression of a ventral valve, with deeply bifurcating costæ. × 1½. Keelbottom River.

Fig. 21. *Chonetes*, sp. ind. Impression of a rather deltoid form. × 2. Keelbottom River.

Fig. 22. *Chonetes*, sp. ind. × 2. Keelbottom River.

PLATE 37.

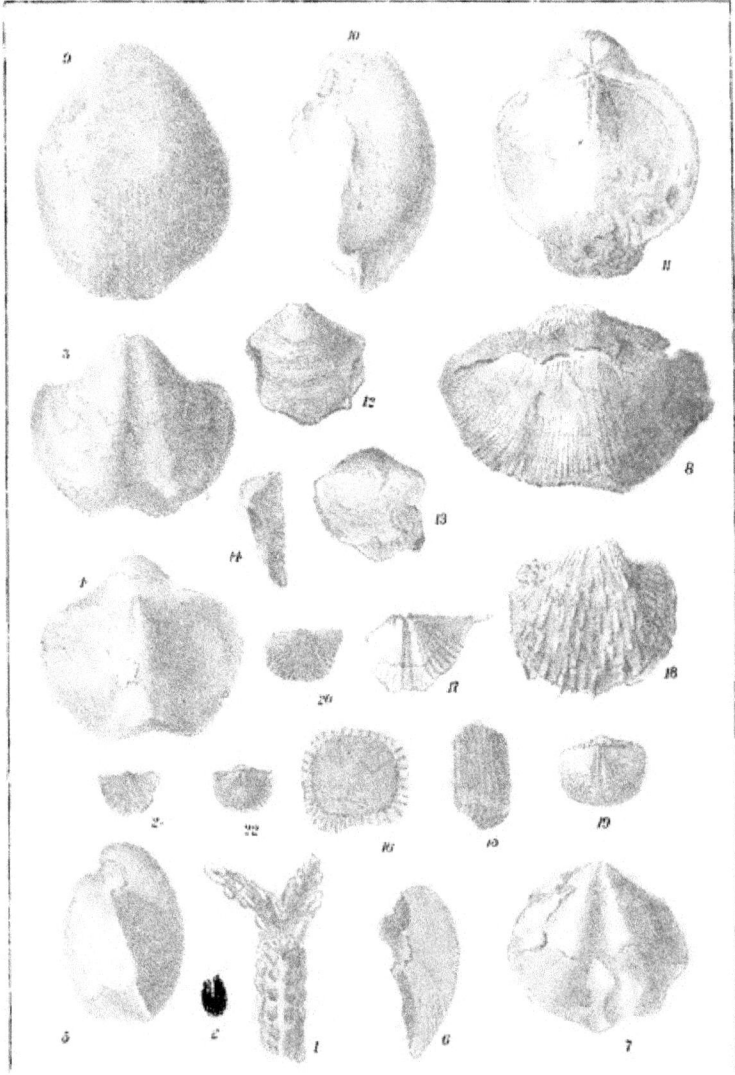

PLATE 38.

Fig. 1. *Monticulipora*, sp. ind.   General feathery-like outline of the corallum.   × 3. Kooingal.

Fig. 2.   Do.   A small portion of the surface, highly enlarged, showing the calices and acanthopores.

Fig. 3. *Platyceras ? sp.*, Eth. fil.   Internal cast of a part of the calyx.   Mount Britton Gold Field.

Fig. 4. *Spirifera Stutchburii*, Eth. fil.   Internal cast of the dorsal valve.   Mount Britton Gold Field.

Fig. 5.   Do.   Ventral valve of the same specimen.

Fig. 6.   Do.   Side view of the united valves.

Fig. 7. *Productus subquadratus*, Morris.   Internal cast of the ventral valve.   Mount Britton Gold Field.

Fig. 8.   Do.   Side view of the same valve.

Fig. 9.   ' Internal cast of the dorsal valve, showing the impressions of the long septum and muscular scars.

Fig. 10.   Do.   Portion of another dorsal valve, with the impressions of the septum and cardinal process extending far into the cavity, the ventral umbo, also the muscular impressions.   Mount Britton Gold Field.

Fig. 11. *Productus cora*, D'Orbigny.   Impression of the exterior of the dorsal valve. Mount Britton Gold Field.

Fig. 12. *Modiomorpha mytiliformis*, Eth. fil.   Decorticated right valve (probably the young condition of Pl. 41, fig. 4).   Mount Britton Gold Field.

Fig. 13.   Do.   Hinge and united valves of the same specimen.

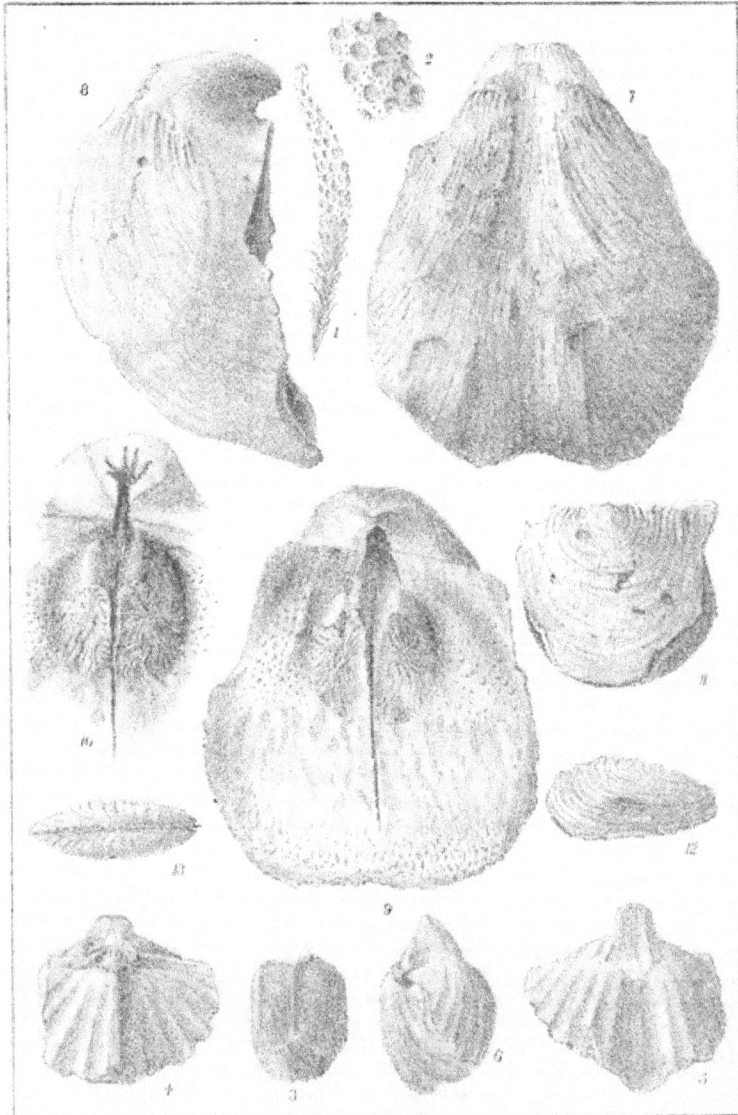

CARBONIFEROUS CORALS, BRACHIOPODA &c.

PLATE 39.

Fig. 1. *Burrow*, with its infilling passing obliquely across the stratification. Rockhampton District (*See* Pl. 8, fig. 4, Pl. 44, figs. 15-8). [De V.]

Fig. 2. *Spirifera Stokesii*, G. B. Sowerby. Internal cast of the dorsal valve. Mount Britton Gold Field.

Fig. 3. Do. Internal cast of the ventral valve of the same specimen.

Fig. 4. Do. Side view, showing the cast of the united valves.

Fig. 5. *Martiaia* (vel. *Martiniopsis*?) *Darwinii*, Morris. Internal cast of the dorsal valve, taken from two specimens, to show the high sharp fold, grooved but undivided, and 2-3 costae on each side. Mount Britton Gold Field.

Fig. 6. Do. Internal cast of the ventral valve.

Fig. 7. Do. Side view of the cast of the united valves of another example, to show the horizontal or depressed umbo of the ventral valve. Mount Britton Gold Field.

Fig. 8. *Undescribed Bivalve* Do. Decorticated right valve. Rockhampton District.

Fig. 9. *Nautilus ammonitiformis*, Eth. fil. The inner whorls (*See* Pl. 41, fig. 9). Rockhampton District. [De V.]

Fig. 10. *Orthoceras*, sp. ind. Cast of five chambers. Mount Britton Gold Field.

Fig. 11. *Deltodus ? australis*, Eth. fil. Rockhampton District. [De V.]

CARBONIFEROUS BRACHIOPODA, &c.

PLATE 40.

Fig. 1. *Dielasma?* sp. ind. Internal cast of the dorsal valve of an imperfect specimen. Rockhampton District. [De V.]

Fig 2. Do. Internal cast of the greater portion of the united valves. [De V.]

Fig. 3. *Spirifera* (allied to *S. oviformis*, McCoy). Internal cast of the dorsal (?) valve. Rockhampton District. [De V.]

Fig. 4. *Productus*, sp. ind. A decorticated example, with a concentrically lined visceral region and a coarsely costate front, somewhat resembling a *Strophomena*. Rockhampton District. [De V.]

Fig. 5. *Productus subquadratus*, Morris. Hinge and interior of the ventral valve, showing a broad area, deltidial opening, and dendritic muscular scars. Yatton Gold Field.

Fig. 6. *Strophomena analoga*, Phillips? Cast of the interior of the ventral valve, with the impressions of the muscular and vascular systems. Rockhampton District. [De V.]

Fig. 7. *Strophalosia Gerardi*, King? Impression of the exterior of a small dorsal valve. Banana Creek. [De V.]

Fig. 8. Do. A ventral valve, with the test preserved, strong concentric laminæ, and blunt recumbent spines. Darr River. [A.L.]

Fig. 9. *Pterinopecten Decisii*, Eth. fil. Right valve. Rockhampton District. [DeV.]

Fig. 10. *Nucula*, sp. ind. Internal cast of the left valve of a very deltoid species, possessing an abrupt anterior end, and a short posterior hinge. Rockhampton District. × 3. [De V.]

Fig. 11. *Mytilops? corrugata*, Eth. fil. Portion of a decorticated valve, with step-like concentric corrugations. × 3. Rockhampton District. [De V.]

Fig. 12. *Parallelodon costellato*, McCoy. Portion of a right valve, decorticated. Rockhampton District. [De V.]

Fig. 13. Do. Internal cast of another right valve, with traces of transverse hinge teeth. Rockhampton District. [De V.]

PLATE 10

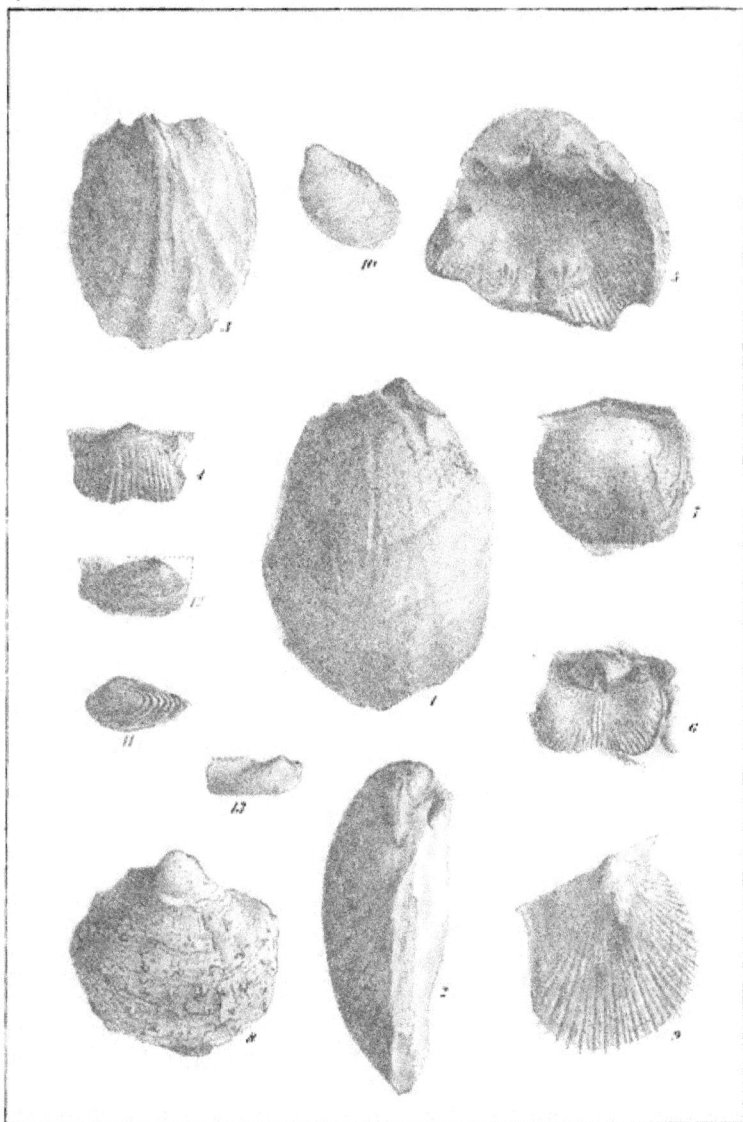

PLATE 41.

Fig. 1. *Lasiocladia? Hindei*, Eth. fil. An impression of the sponge. Rockhampton District. [De V.]

Fig. 2. Do. Portion highly magnified, with impressions of rod-like spicules, and a few *in situ*.

Fig. 3. *Deltopecten illawarrensis*, Morris, sp. Cast of the united valves, to show great inequality. Mount Britton Gold Field.

Fig. 4. *Modiomorpha? mytiliformis*, Eth. fil. Cast of the left valve. Banana Creek (*See* Pl. 14, fig. 5, and Pl. 38, figs. 12 and 13). [De V.]

Fig. 5. *Mourlonia? coniformis*, Eth. fil. Oblique apical view, with coarse reticulated ornament. Banana Creek. [De V.]

Fig. 6. *Luciella? Grayæ*, Eth. fil. Apical view taken from an artificial cast, showing reticular ornament (the nodes at the intersection of the concentric and radiating lines are not represented). × 2. Rockhampton District. [De V.]

Fig. 7. *Yvania Konincki*, Eth. fil. Side view of a partially preserved example. × 2. Rockhampton District. [De V.]

Fig. 8. *Bucania textilis*, De Koninck? Spiral and decussating striæ, both visible. Rockhampton District. [De V.]

Fig. 9. *Nautilus? ammonitiformis*, Eth. fil. More or less complete specimen, with the central small whorls (*See* Pl. 39, fig. 9), and the outer strongly costate whorl. Rockhampton District. [De V.]

Fig. 10. *Conularia*, sp. ind. With transversely wrinkled sutures. Gympie.

Fig. 11. *Goniatites planorbiformis*, Eth. fil. A decorticated example, with the septa visible. Lake's Creek. [De V.]

Fig. 12. *Gyroceras? dubius*, Eth. fil. A fragment seen from the side. Rockhampton District. [De V.]

Fig. 13. *Rhynchonella croydonensis*, Eth. fil. An internal cast of the ventral valve. Croydon.

Fig. 14. Do. Probably the dorsal valve of the same species. Croydon.

PLATE 41

CARBONIFEROUS MOLLUSCA &c

PLATE 42.

Fig. 1. *Phyllotheca*, sp. ind.? Whorl of lanceolate leaves. × 2. Bundanba.

Fig. 2. *Unio ipsviciensis*, Eth. fil. Right valve. Bremer Basin Colliery.

Fig. 3. Do. Hinge and united valves of the same.

Fig. 4. *Maccoyella Barklyi*, var. *mariœburiensis*, Eth. fil. Artificial cast taken from an impression of the left valve. Croydon.

Fig. 5. Do. Small left valve, with spinous terminations to the costæ. Croydon.

Fig. 6. Do. An imperfect right valve with a well-marked lobate ear. Croydon.

Fig. 7. *Inoceramus*, sp. ind. Anterior portion of the united valves. Bowen Downs. —3½ nat. [Q.M.]

Fig. 8. *Rhynchonella*, sp. ind. Impression of a small dorsal valve. Croydon.

Fig. 9. *Siphonaria Samwelli*, Eth. fil. An artifical cast taken from an impression. Croydon.

Fig. 10. *Ammonites walshensis*, Eth. fil. Walsh River. [Q.M.]

Fig. 11. Do. Sectional view of the same specimen.

Fig. 12. *Ammonites*, sp. (allied to *A. denarius*, Sby., or *A. lautus*, Park). An impression with geniculate costæ. Warriana Bore.

Fig. 13. *Ancyloceras? Taylori*, Eth. fil. Head of the Walsh River. [Q.M.]

Fig. 14. *Hamites* (or *Hamulina?*) *laqueus*, Eth. fil. Tower Hill. [Q.M.]

Fig. 15. Do. Section showing a septum from the same specimen.

Fig. 16. *Crioceras irregulare*, Ten. Woods. Walsh River. [Q.M.]

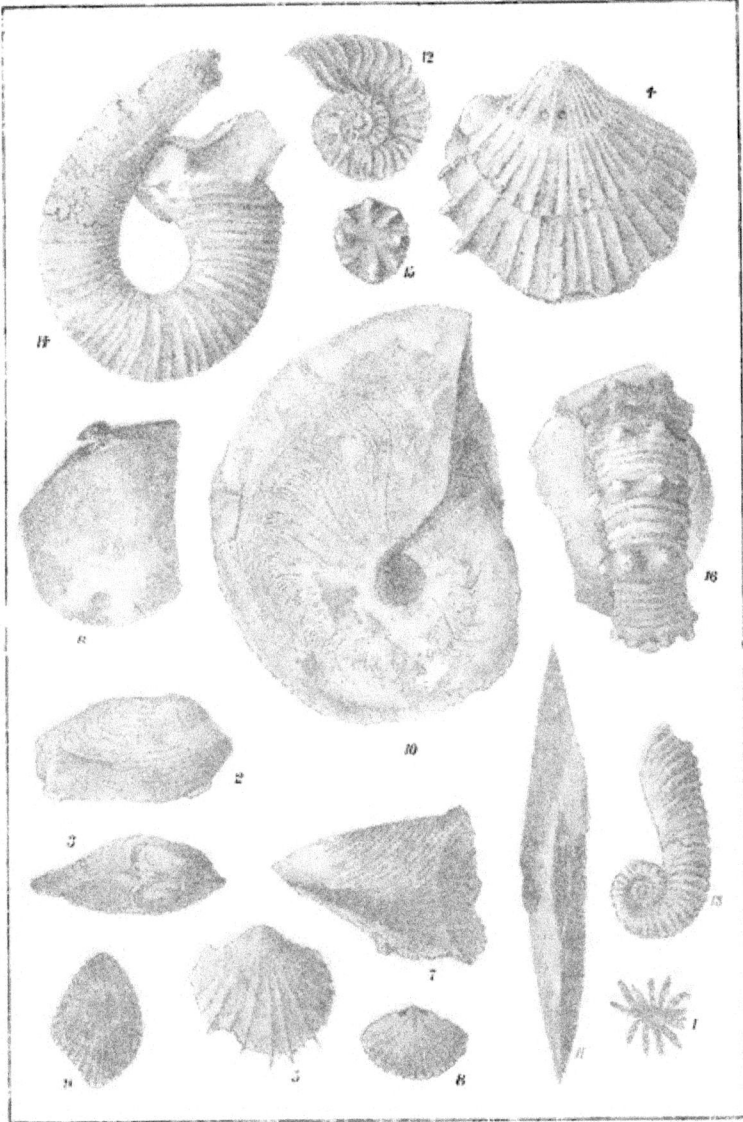

PLATE 43.

Fig. 1. *Martinia* (vel. *Martiniopsis* ?) *subradiata*, G. B. Sby., sp. Internal cast of the united valves, seen from above. Banana Creek. ½ nat. [De V.]

Fig. 2. *Deltopecten illawarrensis*, Morris sp. Cast of a portion of the left (?) valve, with the striated hinge area, and the triangular cartilage-pit below the umbo. Mount Britton Gold Field.

Fig. 3. *Aviculopecten Laurenti*, Eth. fil. Right valve (reversed). Crow's Nest. [M.G.M.]

Fig. 4. Do. Left valve (reversed). Rockhampton District. [De V.]

Fig. 5. *Chænomya* ? *carinata*, Eth. fil. Right valve, showing oblique outline, high umbone and strong diagonal ridge. (This and the next figure have been turned upside down inadvertently.) Banana Creek. [De V.]

Fig. 6. Do. Hinge and united valves. Banana Creek.

Fig. 7. *Sanguinolites concentricus*, Etheridge ? Faint cast of the left valve (reversed). Gympie.

Fig. 8. *Ostrea*, sp. ind. Artificial cast of an impression of the interior of the lower or attached valve. Croydon.

Fig. 9. *Placuna*, sp. ind. Artificial cast of an impression of the interior of the left (?) valve. Croydon.

Fig. 10. *Cardium Brazieri*, Eth. fil. Left (?) valve, with the test partially preserved under its radiating ribs. Loc. ? [J.E.T.W.]

Fig. 11. *Teredo*, sp. ind. Two tubes split in half longitudinally, with the shelly matter partially preserved. Croydon.

Fig. 12. *Teredo*, sp. ind. Tube infilled with matrix, and one valve, shifted from its natural position. ? Loc. [J.E.T.W.]

PLATE 43

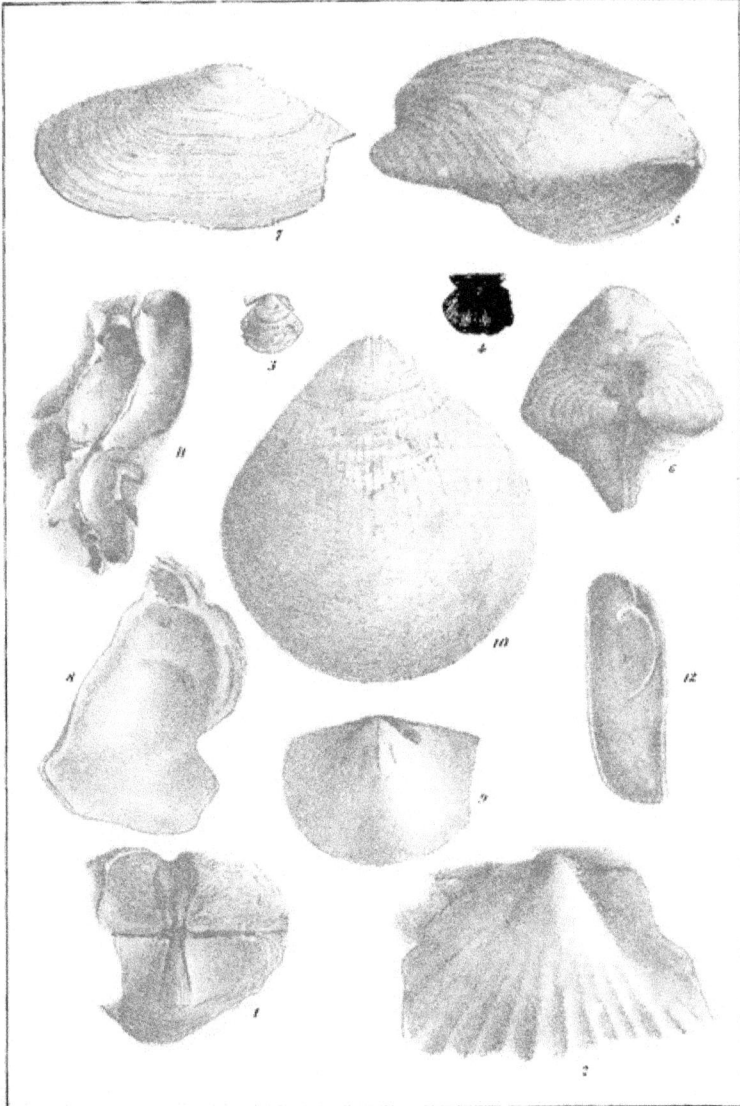

G H Barrow Del et Lith

CARBONIFEROUS & CRETACEOUS

MOLLUSCA

PLATE 44.

Fig. 1. *Zaphrentis profunda*, Eth. fil. Interior of the calice. Rockhampton District. [De V.]

Fig. 2. *Mesoblastus ? australis*, Eth. fil. Side view of an impression of a calyx. × ½. Rockhampton District. [De V.]

Fig. 3. *Tricœlocrinus Carpenteri*, Eth. fil. Side view of an internal cast. Rockhampton District. [De V.]

Fig. 4. *Phillipsia dubia*, Etheridge. Thorax and pygidium decorticated. Rockhampton District. [De V.]

Fig. 5. *Phillipsia Woodwardi*, Eth. fil. Glabella with large basal lobes. × 2. Rockhampton District. [De V.]

Fig. 6. Do. Pygidium probably of this species. × ½. Rockhampton District. [De V.]

Fig. 7. *Crinoid* (allied to *Stemmatocrinus*). Basal cup denuded of the test. Rockhampton District. [De V.]

Fig. 8. *Crinoid Calyx*. Impression of part of an indistinct calyx, with faint traces of plates and the bases of the arms. Rockhampton District. [De V.]

Fig. 9. *Polypora ? Smithii*, Eth. fil. Fragment of a polyzoarium. Rockhampton District. × 2.

Fig. 10. Do. Fragment of poriferous face greatly enlarged.

Fig. 11. *Glauconome* sp. ind. Non-poriferous face. × 2. Rockhampton District. [De V.]

Fig. 12. *Spiriferina duodecimcostata*, McCoy. Internal cast of the ventral valve, showing the slit left by the median ventral septum, and the casts of the punctæ of the shell foramina. Mount Britton Gold Field. [A.M.]

Fig. 13. *Productus*, sp. ind. Much resembling the young of *P. giganteus*, Martin. Yatton Gold Field.

Fig. 14. *Productus brachythærus*, G. B. Sowerby. Showing an area, and considerable solidity of shell. Darr River ? [A.L.]

Fig. 15. *Burrow*. Weathered and disjointed infillings of the burrow. (*See* Pl. viii., fig. 4, and Pl. xxix., fig. 1.) Rockhampton District. [De V.]

Fig. 16. Do. A similar specimen. [De V.]

Fig. 17. Do. Another example seen in natural section. [De V.]

Fig. 18. Do. Transverse section displaying the involute structure of the burrows. × 3. Rockhampton District. [De V.]

By Authority: JAMES C. BEAL, Government Printer, William street, Brisbane.

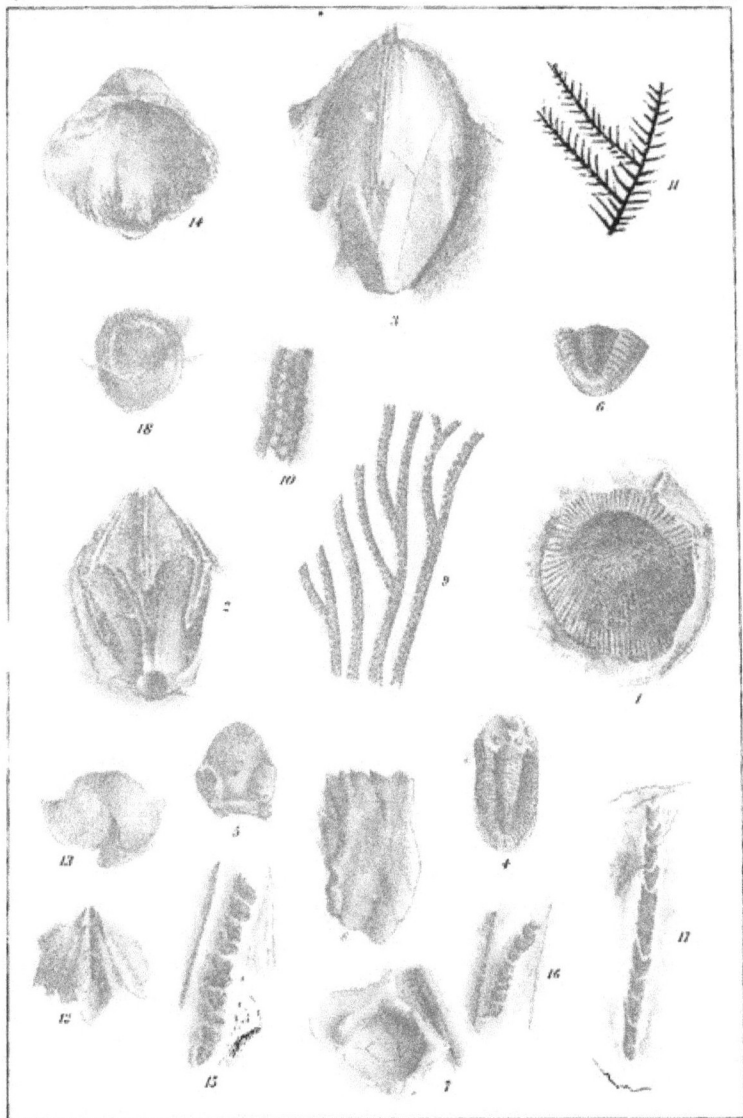

G. H. Barrow del et lith

CARBONIFEROUS MISCELLANEA

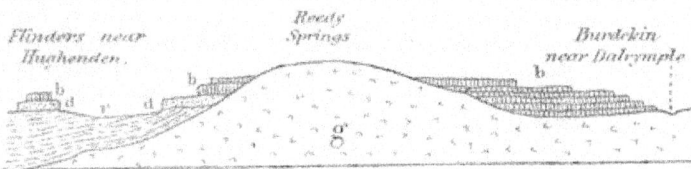

Fig. 1. Diagram—Section from Dalrymple to the Head of
the Flinders.

b. Basalt. d. Desert Sandstone. r Rolling Downs Formation. g Granite

Fig. 2. Diagram—Section across
True Blue Hill, Croydon Goldfield.

d. Desert Sandstone. d'. The same un-
consolidated, forming sandy flats. g. Granite.

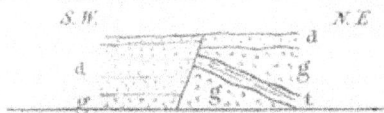

Fig. 3. Section in True Blue 3 & 4 South, Croydon.

d. Desert Sandstone. g. Granite. t. True Blue Reef.

Fig.1. Section across Mt Mandarana, near Mackay.

S. Carbonifero-Permian Sandstones and Shales.   B. Basalt.
V. Volcanic Series. T. Trachyte Lava.   T₁. Trachyte-Ash.

Fig.2. Section across the Finlayson Hills, near Mackay.
G. Granite.   T. Trachyte Lava.

Fig.3. Diagram—Section showing supposed relation
of 'Maryborough Beds' to the Burrum Coalfield.

D Maryborough Beds. T. Tinana Coal Series. B.Burrum Coal Series
f. Fault.

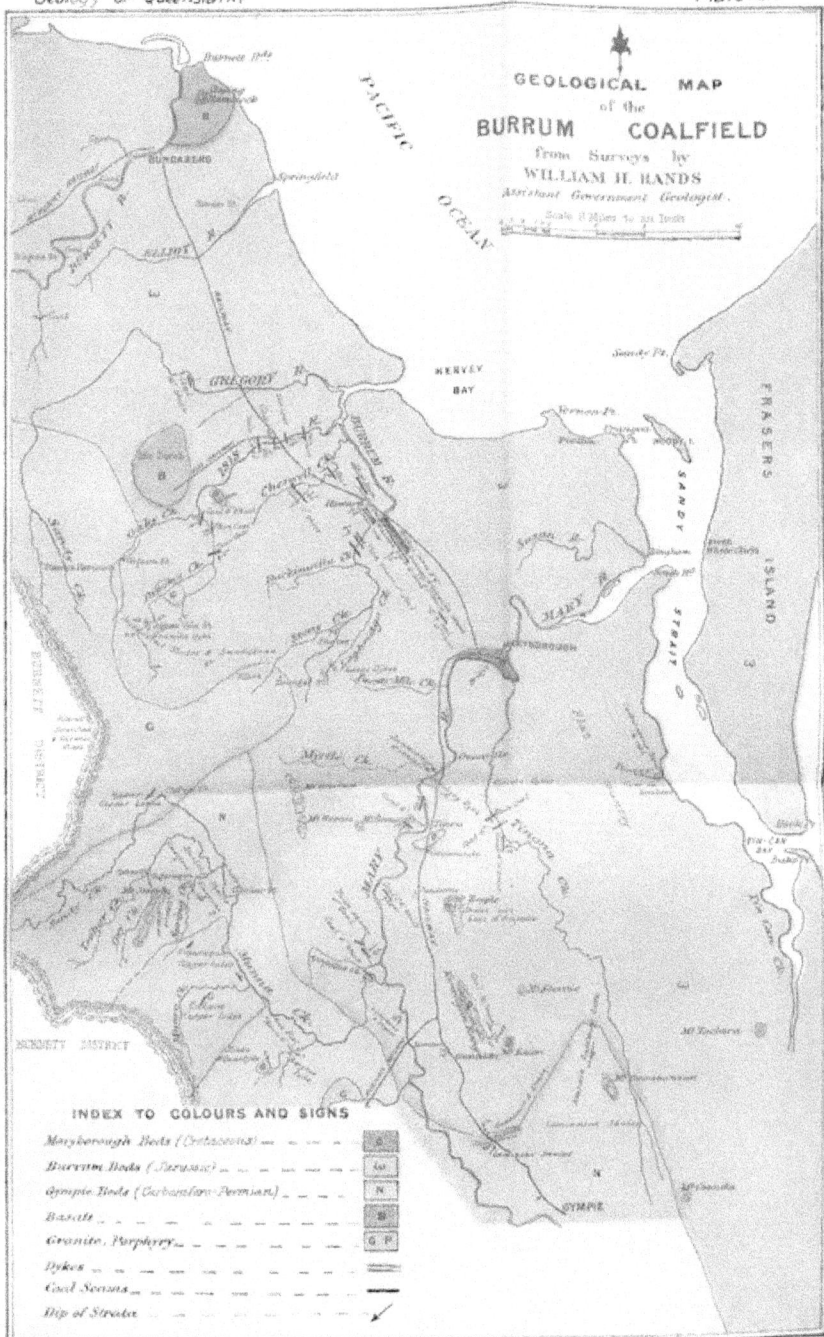

GEOLOGICAL   MAP
of the
BURRUM   COALFIELD
from Surveys by
WILLIAM H. RANDS
Assistant Government Geologist.

Scale 8 Miles to an Inch

INDEX TO COLOURS AND SIGNS

Maryborough Beds (Cretaceous) _ _ _   M
Burrum Beds (Jurassic) _ _ _ _   w
Gympie Beds (Carboniferous-Permian) _ _ _   N
Basalt _ _ _ _ _   B
Granite, Porphyry _ _ _ _   G P
Dykes _ _ _ _ _
Coal Seams _ _ _ _
Dip of Strata _ _ _ _

PRINTED AT THE GOVT. ENGRAVING & LITHOGRAPHIC OFFICE, BRISBANE, QUEENSLAND.
W. KNIGHT, LITH. ENGRAVER.

GEOLOGICAL MAP

of the

GYMPIE GOLD FIELD

from Survey by

WILLIAM H. RANDS

Assistant Government Geologist

1893

Scale 40 Chains to an Inch

INDEX OF COLOURS AND SIGNS

Limestone	*Shaley & Slate bedded Sandstone*
Slate of the Gold Slate bed of Gympie	*Green ish Sandstone or Serpentine*
Conglomerate	*Greenstone*
Old Alluvium	*Recent Alluvium*
Old Alluvial workings	*Igneous Dykes & Sheets*
Faults (triangle shows direction of downthrow)	
Reefs (triangle shows direction of underlie)	
Dip of Strata (number denotes amount of dip in degrees)	

Fig. 1.

N.S   S.W

Section across Limestone Hill Ipswich

a. Limestone.   b. Shales.   c. Basalt.

Fig 2.

ROSEWOOD SCRUB

W   IPSWICH   BRISBANE   E

Line 1.

TOOWOOMBA RANGE

W   E

Line 2.

WESTERN DOWNS   CLIFTON

W   E

Line 3.

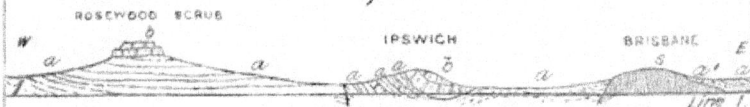

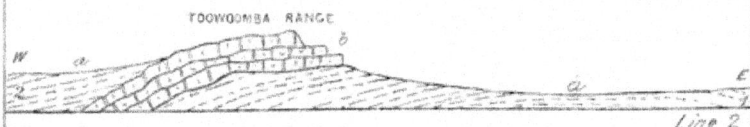

Diagram Section, Brisbane to the Downs

a'. Ashy Sandstones. a. Ipswich Formation. b. Basalts.

c. Rolling Downs Formation.   d. Desert Sandstone

S Palæozoic Slates, Quartzites &c

Fig. 3.

Section in Railway Cutting near Walton Station

a. Shales (Ipswich Formation).   b. Sanidine Trachyte.

Fig. 1.

NW

SE | N   S

Ben Lomond
(Star Beds)   Camp Hill
(Star Beds)

Little Star River   Argentine
Star Stn.   Silver Field

Star Beds.   Granite.   Slates & Schists
(Type District)

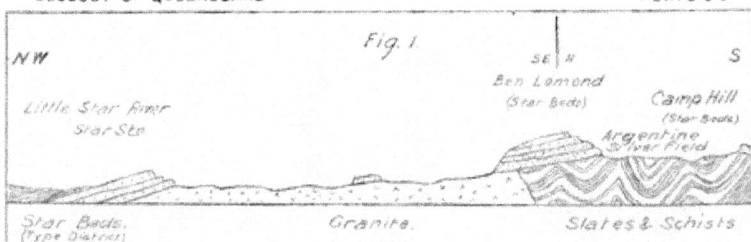

Section from Little Star River to Argentine Silver Field.

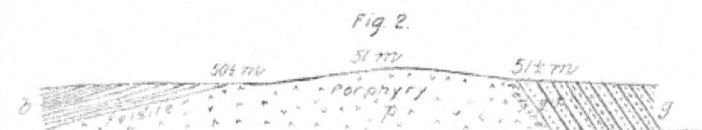

Fig. 2.

504 m.   51 m.   51½ m.

b   Porphyry   g
Felsite   p.

Section in Railway Cutting S. of Curra showing
relations of Burrum and Gympie Beds

b. Burrum Beds   g. Gympie Beds   p. Porphyry & Felsite

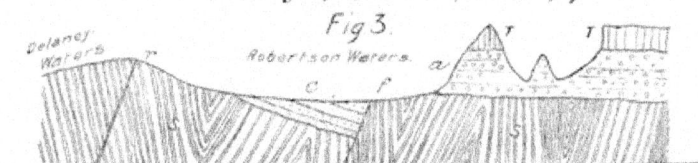

Fig. 3.

Delaney
Waters   T'   Robertson Waters   a   T   T
                                   c   f

Section from Goldsmith's Reefs (Delaney Waters)
Southward across the Heads of the Robertson

T Bedded Trachyte   a. Coarse Ash passing into Conglomerate
c. Star Beds. S, Schists &c   f Fault   r. Lady Franklin.

Fig. 3.

Fig. 1.

Fig. 2.

Diamond Drill Core

No. Pelican Co. Bore. Bowden River at 191½. Showing "White Trap" intruding into base of a Coal Seam. ½ Natural size.

Diamond Drill Core

Same Bore at 507½ 9 inches Showing masses of "White Trap" in burnt Coal Seam.

Diamond Drill Core

Same Bore at 288½ ½. Showing "White Trap" intruding into Coal Seam, with threads of Coal penetrating the "White Trap" ½ of Natural size.

Fig. 4.

Coal Seam destroyed by intrusive Sheet of Dolerite. No. 2 Travers Station, Rosella Creek.

g. Burnt Coal, 3 inches. a. Dolerite- "White Trap" in part, involving angular blocks of Coal, b. Burnt Coal, upper part columnar 3½ 6 inches. c. Blue Shale, topmost part hardened. 2½ 1 inch. d. Burnt Coal. e. Black Clay, 3 inches. f. Burnt Coal, 3 feet.

Fig. 5.

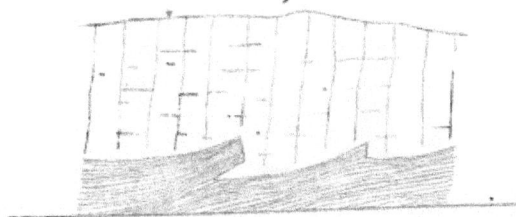

Vertical Section, showing Junction of Shales and intrusive Dolerite Sheet. Rosella Ck.

Fig. 1.

Vertical Section of Shales, with "Pocket" of Mineral Pitch.
Rosella Creek

Fig. 2.

Top of   Bank
                                                            b
                                                            a
Level of   Jacks Creek

a. Shales and Sandstone. b Dolerite.
Section near Mouth of Jack's Creek

Fig. 3

Capping of Dolerite on top of bank

Sloping grassy bank

Gully shewing at least 5 ft more of Coal, mixed with white trap.

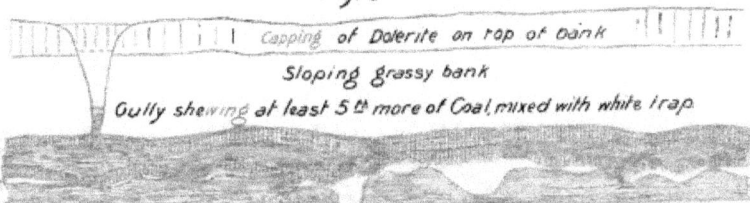

Havilah Coal Seam, Rosella Creek, burned and partly rendered
Columnar by White Trap.

Fig. 4.

N
W   E
S

Dark shales
one layer coarse

Dark shales
& hardened sandstone bands
(on end)

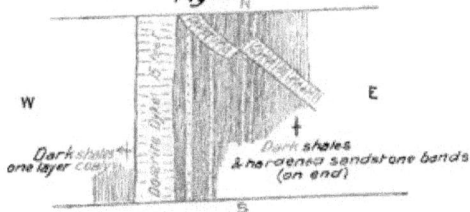

Plan of Bed of Jacks Creek, Bowen River, shewing Dolerite Dykes
intersecting Shales and Sandstone

Fig. 1.

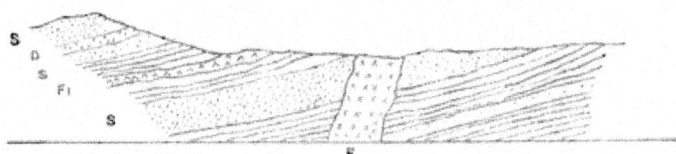

Section on Alligator Creek near Mackay

S Sandstone and Shales (Lower Series, Bowen River Beds)
D Diorite Bed (contemporaneous)
F₁ Felstone Sheet (intrusive)
E Felsite Dyke

Fig 2.

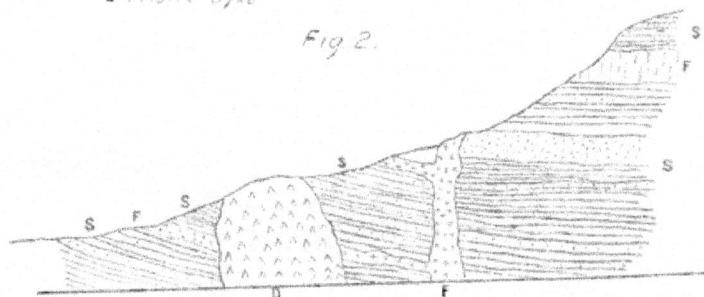

Section in Bells Creek near Mackay

S Grits and Shales (Lower Series, Bowen River Coalfield)
F Intrusive Felsite (in dyks and sheets)
D Intrusive Diorite

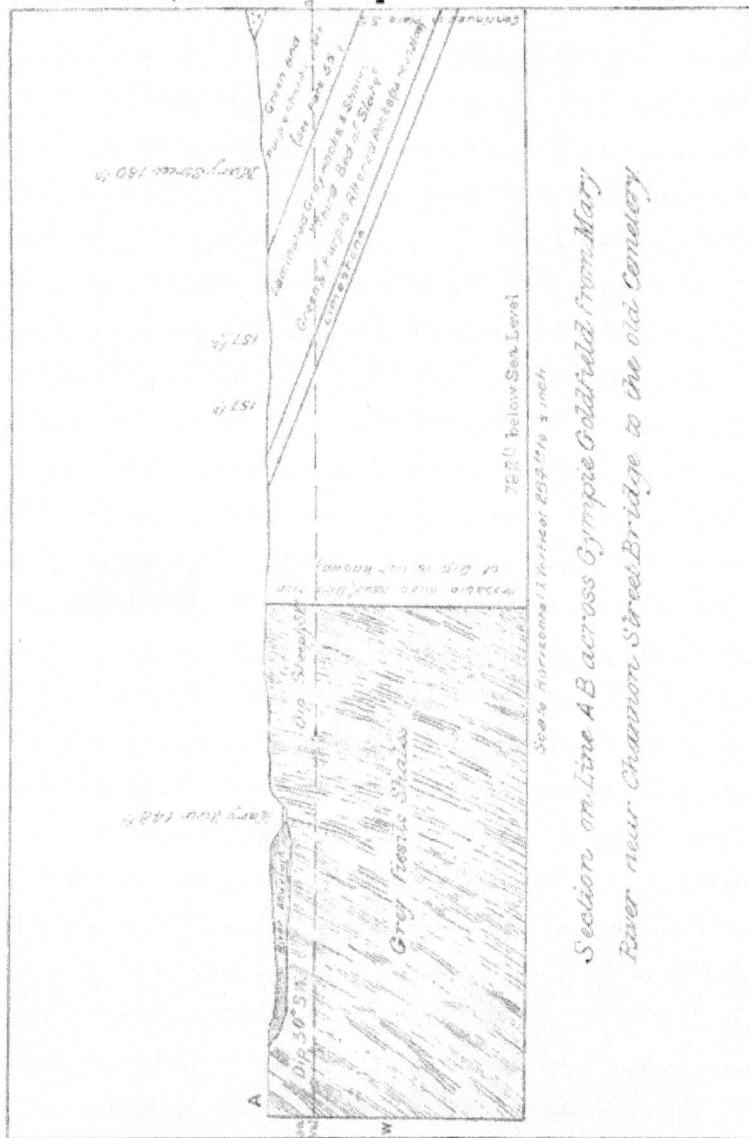

Section on Line AB across Gympie Goldfield from Mary River near Channon Street Bridge to the old Cemetery.

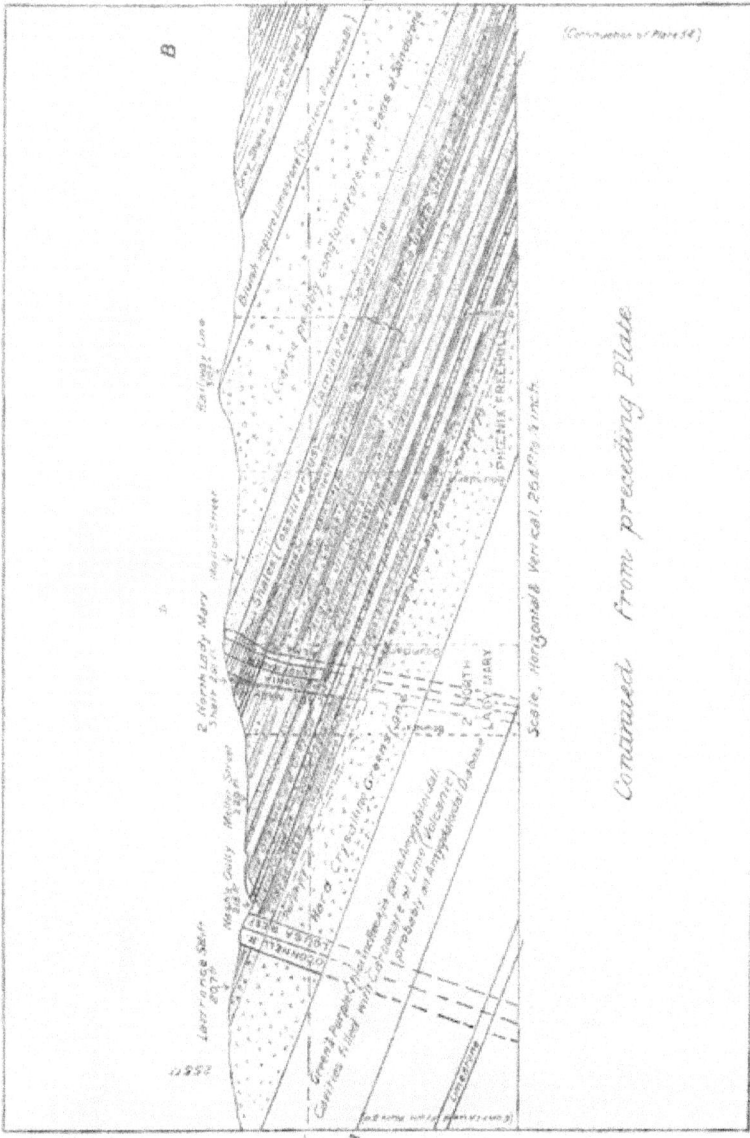

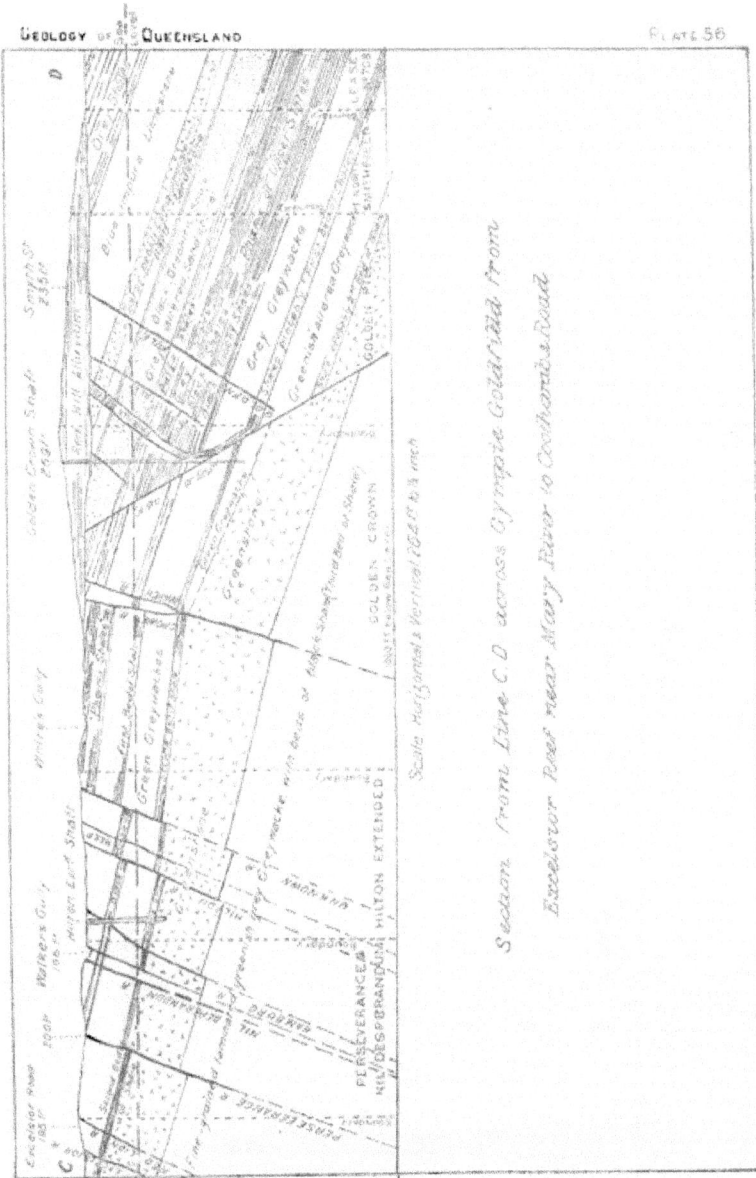

Section from Line C.D across Gympie Goldfield from
Excelsior Road near Mary River to Coolhoma Road

Scale Horizontal & Vertical 704ft to ½ inch

Smyth St
735ft

Golden Crown Shaft
743ft

Phillips Gully

Watkins Gully

Hilton Ext. Shaft

Excelsior Road
185ft

GOLDEN CROWN

GOLDEN PIKE

Grey Greywacke

Green Greywacke

PERSEVERANCE
NIL/DESP/BRANDUM HILTON EXTENDED
PERSEVERANCE

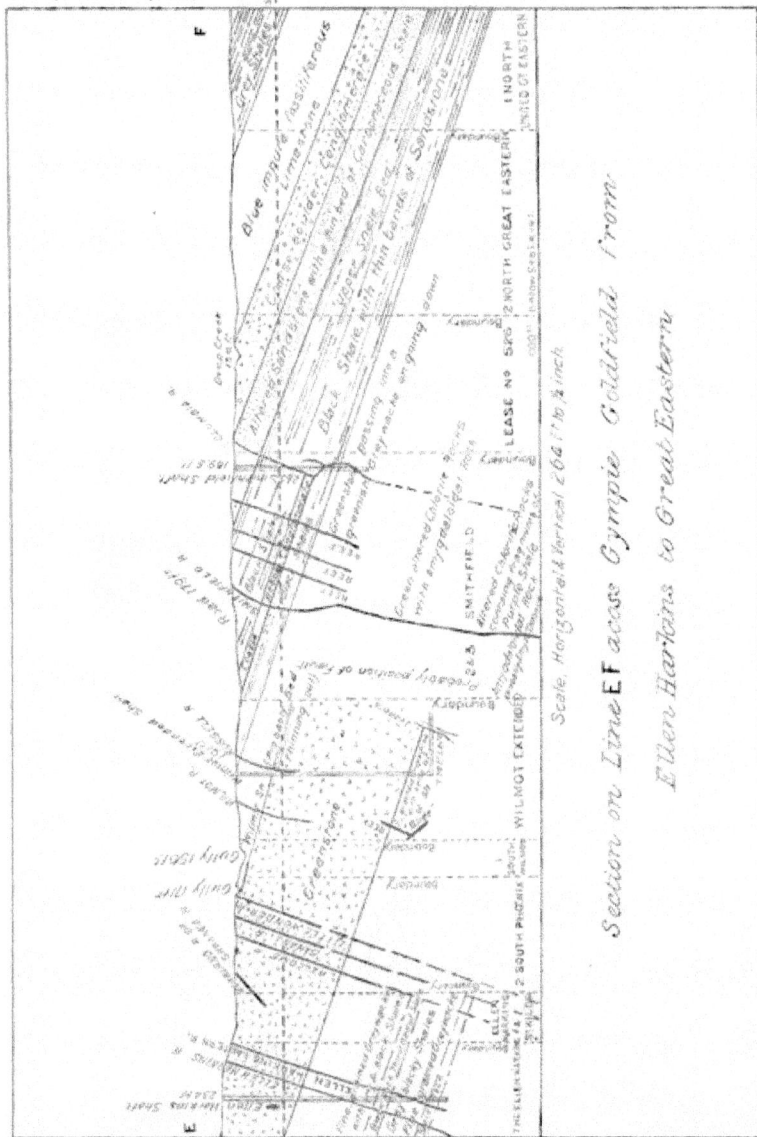

Section on Line EF across Gympie Goldfield from Ellen Harkins to Great Eastern

Scale. Horizontal & Vertical 264 ft to ½ inch

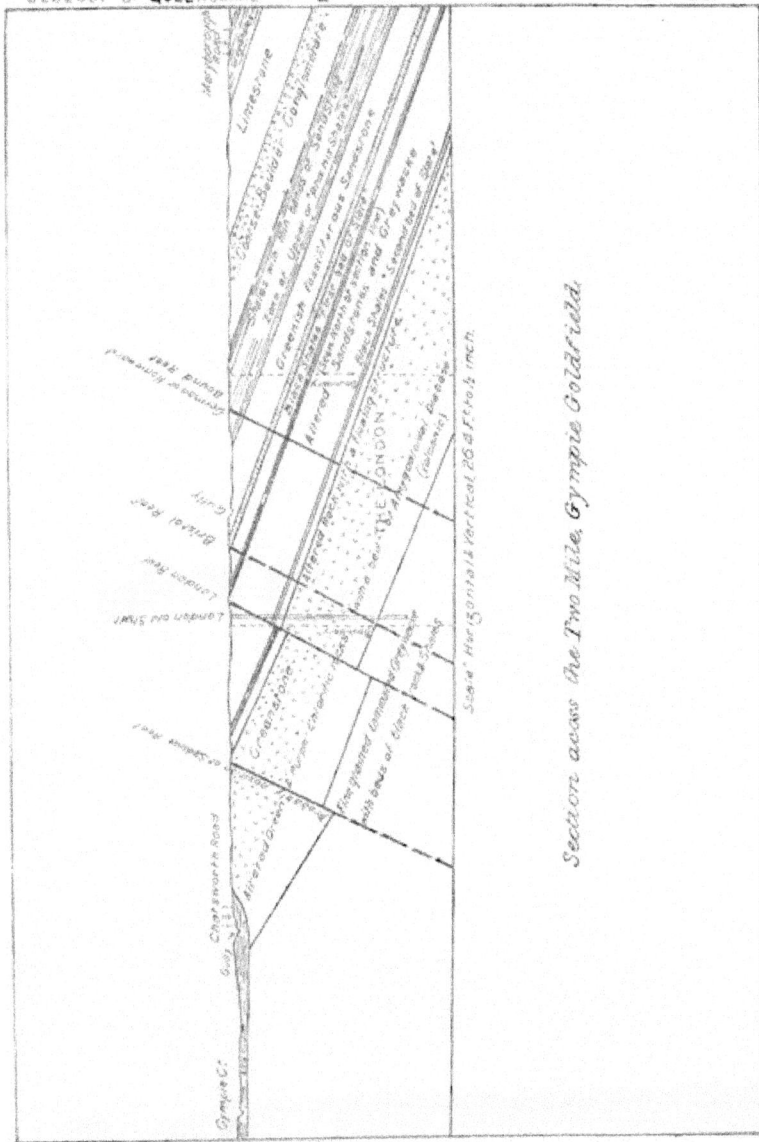

Section across the Two Mile, Gympie Goldfield.

Scale: Horizontal 1 Vertical 26.4 Ft to 1 inch.

Limestone

Coarse Beds of Conglomerate

Beds with thin Bands of Sandstone

Greenish Reddish Ferruginous Sandstone

Black Shales and Greywacke

THE LONDON

Greenstone

W

W

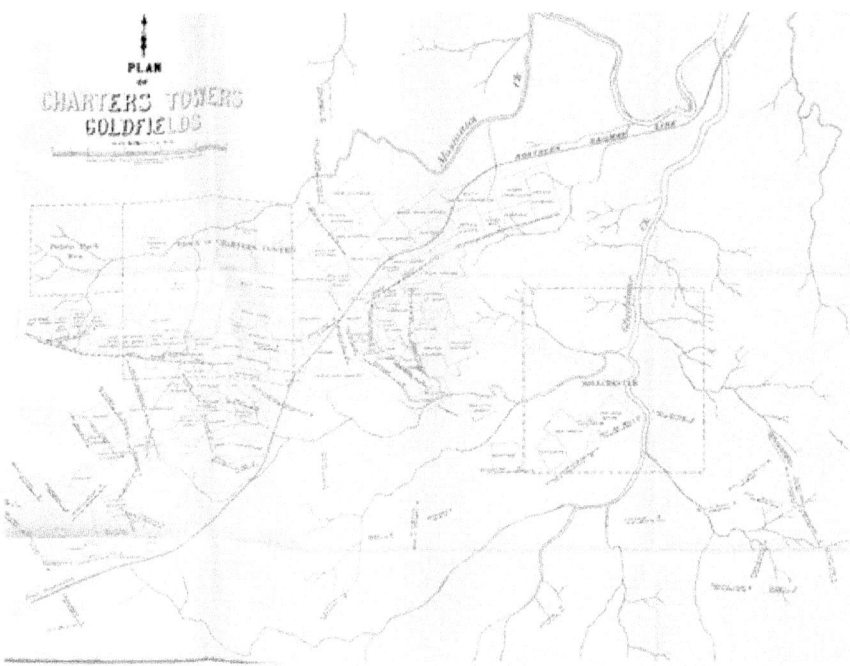

PLAN
OF
CHARTERS TOWERS
GOLDFIELDS

INTRUSIVE DYKE.
Peak, N.S.W.
x50 (With polarizer ends)

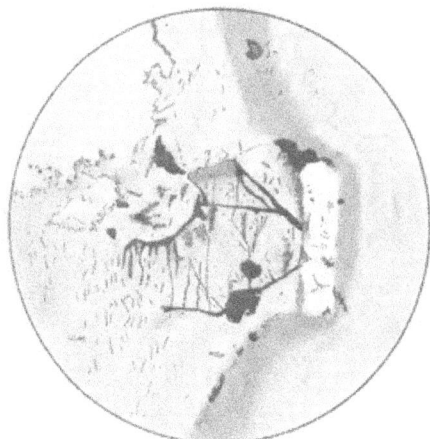

Fig.1.
RHYOLITE
Nell 1ᵈ New Guinea.
142(Polarizer only.)

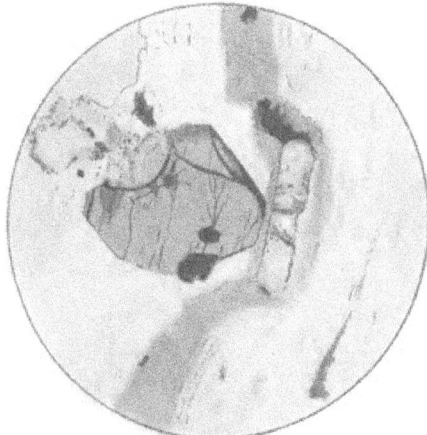

Fig. 2.
RHYOLITE
Nell 1ᵈ New Guinea.
142(Parallel nicols)

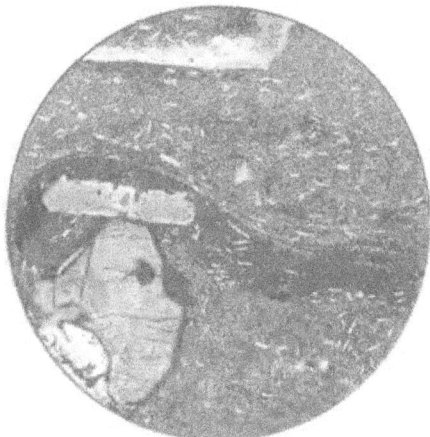

Fig. 1.
RHYOLITE.
Neil 1st New Guinea. ×42 (Crossed nicols)

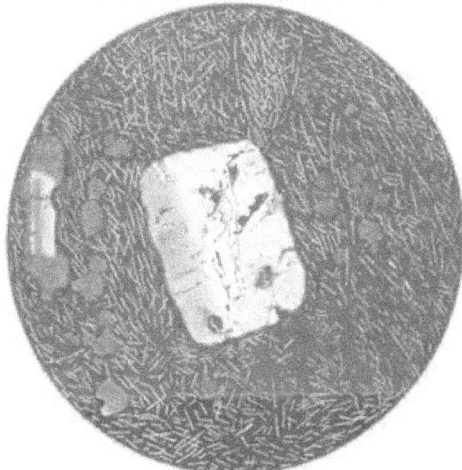

Fig. 2.
RHYOLITE.
Mackay
×275 (Crossed nicols.)

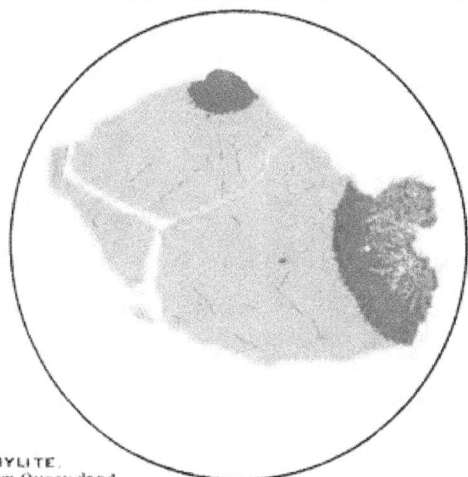

TACHYLITE.
Mitchell River, Queensland.
X27.5 (Polarizer only.)

Fig 1.

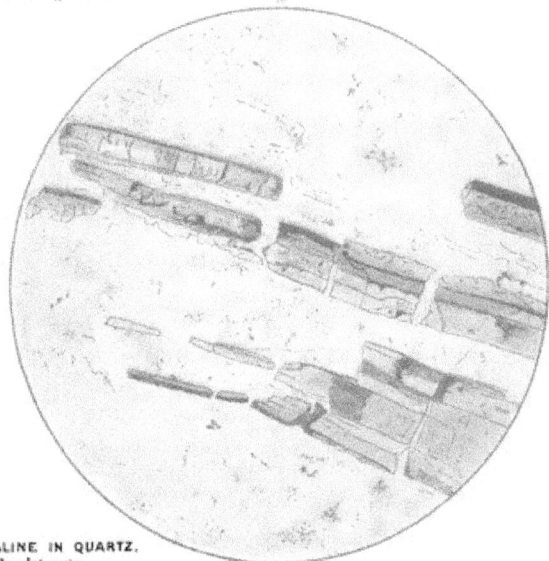

TOURMALINE IN QUARTZ.
Cooktown.
142 (Polarizer only.)

Fig. 2.

Fig. 1.

Capel'pstart.
275 (Crossed nicols.)

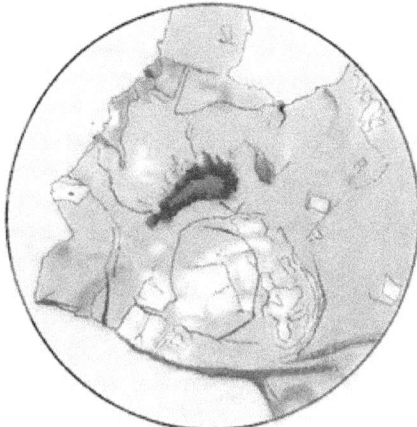

Fig. 2.
PITCHSTONE
Sheffield Tasmania.
142(Polarizer only.)

A. A. Wright lith.   DRAWN & PRINTED AT THE SURVEYOR GENERALS OFFICE, BRISBANE 1892.   J. Phœbe Clarke, del.

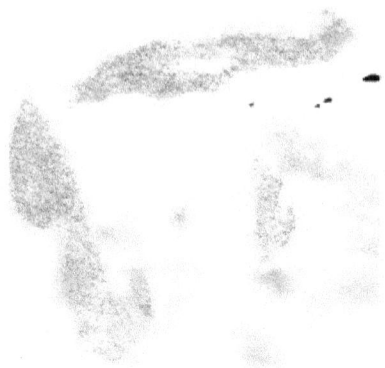

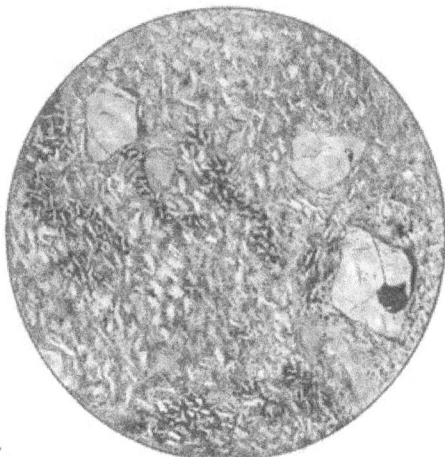

BASALT.
Russell River.
150 (Crossed nicols.)

Fig. 1.

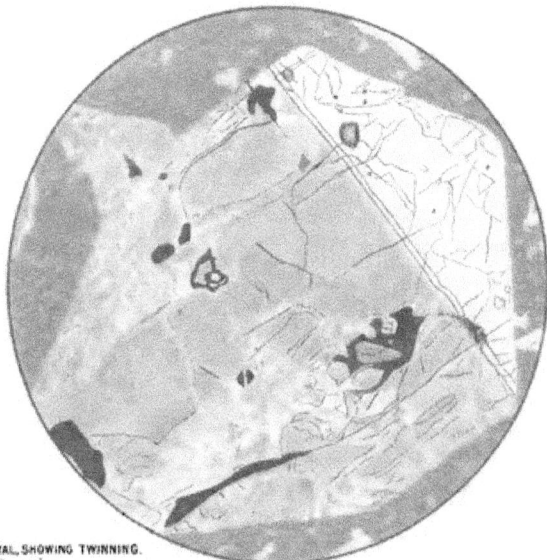

AUGITE CRYSTAL, SHOWING TWINNING.
Gympie.
150(crossed nicols.)

Fig. 2.

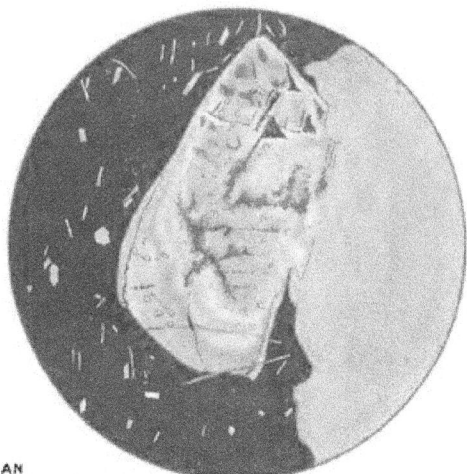

OBSIDIAN
Fergusson Id New Guinea.
150 (Crossed nicols.)

Fig. 1.

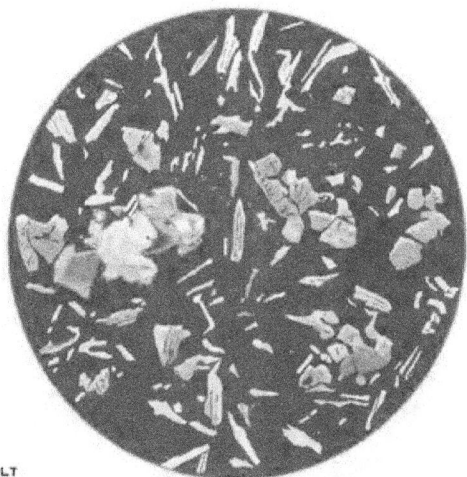

BASALT
Burdekin River.
127.5 (Crossed nicols.)

Fig 2.

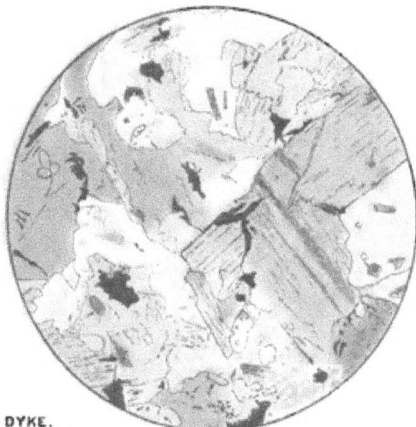

HORNBLENDIC DYKE.
Durham Mine, Etheridge.
275(Polarizer only.)   Fig 1.

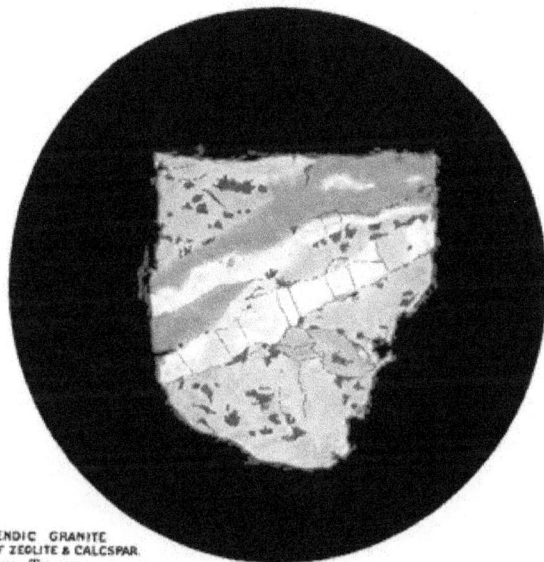

HORNBLENDIC GRANITE
WITH VEINS OF ZEOLITE & CALCSPAR.
Charters Towers.
Natural size. Polished face.   Fig. 2.

A. A. Wright, lith.   DRAWN & PRINTED AT THE SURVEYOR GENERALS OFFICE, BRISBANE 1892   J. Phœbe Clarke del.

Daunton'sHill, Upper Cape.
142(Crossed nicols.)

A. A. Wright, lith.   DRAWN & PRINTED AT THE SURVEYOR GENERALS OFFICE BRISBANE 1892.   J. Photo Clarke, del.